Newnes Radio and RF Engineering
Pocket Book

Newnes Radio and RF Engineering Pocket Book

3rd edition

Steve Winder
Joe Carr

Newnes

OXFORD   AMSTERDAM   BOSTON   LONDON   NEW YORK   PARIS
SAN DIEGO   SAN FRANCISCO   SINGAPORE   SYDNEY   TOKYO

Newnes
An imprint of Elsevier Science
Linacre House, Jordan Hill, Oxford OX2 8DP
225 Wildwood Avenue, Woburn, MA 01801-2041

First published 1994
Reprinted 2000, 2001
Second edition 2000
Third edition 2002

British Library Cataloguing in Publication Data
A catalogue record for this book is available from the British Library

ISBN 0 7506 5608 5

For information on all Newnes publications
visit our website at www.newnespress.com

Typeset by Laserwords Private Limited, Chennai, India.
Transferred to digital printing 2006
Printed and bound by CPI Antony Rowe, Eastbourne

Contents

Preface to second edition

This edition of the *Newnes Radio and RF Engineer's Pocket Book* is something special. It is a compendium of information of use to engineers and technologists who are engaged in radio and RF engineering. It has been updated to reflect the changing interests of those communities, and reflects a view of the technology like no other. It is packed with information!

This whole series of books is rather amazing with regard to the range and quality of the information they provide, and this book is no different. It covers topics as diverse as circuit symbols and the abbreviations used for transistors, as well as more complex things as satellite communications and television channels for multiple countries in the English speaking world. It is a truly amazing work.

We hope that you will refer to this book frequently, and will enjoy it as much as we did in preparing it.

John Davies
Joseph J. Carr

Acknowledgements

I gratefully acknowledge the ready assistance offered by the following organizations: Andrew Ltd, Aspen Electronics Ltd, BBC, British Telecommunications plc, Farnell Instruments Ltd, Independent Television Authority, International Quartz Devices Ltd, Jaybeam Ltd, MACOM Greenpar Ltd, Marconi Instruments Ltd, Panorama Antennas Ltd, Radiocommunications Agency, the Radio Authority, RTT Systems Ltd. A special thanks goes to my wife Dorothy for once again putting up with my months of seclusion during the book's preparation.

Preface to third edition

This, the third edition of the *Newnes Radio and RF Engineering Pocket Book* has been prepared with a tinge of sadness. Joe Carr, who edited the second edition, has died since the last edition was published. Although I did not know Joe personally, his prolific writing over recent years has impressed me. His was a hard act to follow.

I have updated this book to be more international. Thus the long tables giving details of British television transmitters have been removed (they are available on the Web). Details of the European E1 multiplexing system have been supplemented by a description of the US and Japanese T1 system. There are many more general updates included throughout.

Steve Winder

1 Propagation of radio waves

1.1 Frequency and wavelength

There is a fixed relationship between the frequency and the wavelength, which is the distance between identical points on two adjacent waves (*Figure 1.1*), of any type of wave: sound (pressure), electromagnetic (radio) and light. The type of wave and the speed at which the wavefront travels through the medium determines the relationship. The speed of propagation is slower in higher density media.

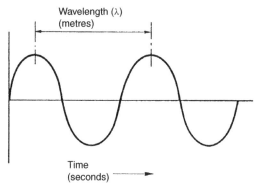

Figure 1.1 Frequency and wavelength

Sound waves travel more slowly than radio and light waves which, in free space, travel at the same speed, approximately 3×10^8 metres per second, and the relationship between the frequency and wavelength of a radio wave is given by:

$$\lambda = \frac{3 \times 10^8}{f} \text{ metres}$$

where λ is the wavelength and f is the frequency in hertz (Hz).

1.2 The radio frequency spectrum

The electromagnetic wave spectrum is shown in *Figure 1.2*: the part usable for radio communication ranges from below 10 kHz to over 100 GHz.

Figure 1.2 The electromagnetic wave spectrum

The radio spectrum is divided into bands and the designation of the bands, their principal use and method of propagation is shown in *Table 1.1*. Waves of different frequencies behave differently and this, along with the amount of spectrum available in terms of radio communication channels in each band, governs their use.

Table 1.1 Use of radio frequencies

Frequency band	Designation, use and propagation
3–30 kHz	Very low frequency (VLF). Worldwide and long distance communications. Navigation. Submarine communications. Surface wave.
30–300 kHz	Low frequency (LF). Long distance communications, time and frequency standard stations, long-wave broadcasting. Ground wave.
300–3000 kHz	Medium frequency (MF) or medium wave (MW). Medium-wave local and regional broadcasting. Marine communications. Ground wave.
3–30 MHz	High frequency (HF). 'Short-wave' bands. Long distance communications and short-wave broadcasting. Ionospheric sky wave.
30–300 MHz	Very high frequency (VHF). Short range and mobile communications, television and FM broadcasting. Sound broadcasting. Space wave.
300–3000 MHz	Ultra high frequency (UHF). Short range and mobile communications. Television broadcasting. Point-to-point links. Space wave. Note: The usual practice in the USA is to designate 300–1000 MHz as 'UHF' and above 1000 MHz as 'microwaves'.
3–30 GHz	Microwave or super high frequency (SHF). Point-to-point links, radar, satellite communications. Space wave.
Above 30 GHz	Extra high frequency (EHF). Inter-satellite and micro-cellular radio-telephone. Space wave.

1.3 The isotropic radiator

A starting point for considering the propagation of radio- or lightwaves
is the isotropic radiator, an imaginary point source radiating equally
in all directions in free space. Such a radiator placed at the centre of
a sphere illuminates equally the complete surface of the sphere. As
the surface area of a sphere is given by $4\pi r^2$ where r is the radius of
the sphere, the brilliance of illumination at any point on the surface
varies inversely with the distance from the radiator. In radio terms,
the power density at distance from the source is given by:

$$P_d = \frac{P_t}{4\pi r^2}$$

where P_t = transmitted power.

1.4 Formation of radio waves

Radio waves are electromagnetic. They contain both electric and magnetic fields at right angles to each other and also at right angles to
the direction of propagation. An alternating current flowing in a conductor produces an alternating magnetic field surrounding it and an
alternating voltage gradient – an electric field – along the length of
the conductor. The fields combine to radiate from the conductor as in
Figure 1.3.

Figure 1.3 Formation of electromagnetic wave

The plane of the electric field is referred to as the E plane and that
of the magnetic field as the H plane. The two fields are equivalent to
the voltage and current in a wired circuit. They are measured in similar
terms, volts per metre and amperes per metre, and the medium through

which they propagate possesses an impedance. Where $E = ZI$ in a wired circuit, for an electromagnetic wave:

$$E = ZH$$

where
E = the RMS value of the electric field strength, V/metre
H = the RMS value of the magnetic field strength, A/metre
Z = characteristic impedance of the medium, ohms

The voltage is that which the wave, passing at the speed of light, would induce in a conductor one metre long.

The characteristic impedance of the medium depends on its permeability (equivalent of inductance) and permittivity (equivalent of capacitance). Taking the accepted figures for free space as:

$\mu = 4\pi \times 10^{-7}$ henrys (H) per metre (permeability) and

$\varepsilon = 1/36\pi \times 10^9$ farads (F) per metre (permittivity)

then the impedance of free space, Z, is given by:

$$\sqrt{\frac{\mu}{\varepsilon}} = 120\pi = 377 \, \text{ohms}$$

The relationship between power, voltage and impedance is also the same for electromagnetic waves as for electrical circuits, $W = E^2/Z$.

The simplest practical radiator is the elementary doublet formed by opening out the ends of a pair of wires. For theoretical considerations the length of the radiating portions of the wires is made very short in relation to the wavelength of the applied current to ensure uniform current distribution throughout their length. For practical applications the length of the radiating elements is one half-wavelength ($\lambda/2$) and the doublet then becomes a dipole antenna (*Figure 1.4*).

When radiation occurs from a doublet the wave is polarized. The electric field lies along the length of the radiator (the E plane) and the magnetic field (the H plane) at right angles to it. If the E plane is vertical, the radiated field is said to vertically polarized. Reference to the E and H planes avoids confusion when discussing the polarization of an antenna.

Unlike the isotropic radiator, the dipole possesses directivity, concentrating the energy in the H plane at the expense of the E plane. It effectively provides a power gain in the direction of the H plane

--- --- Voltage distribution
——— Current distribution

Figure 1.4 Doublet (dipole) antenna

Figure 1.5 Free space loss vs. distance and frequency

Figure 1.6 Major loss of microwave communications and radar systems due to atmospheric attenuation

compared with an isotropic radiator. This gain is 1.6 times or 2.15 dBi (dBi means dB relative to an isotropic radiator).

For a direct ray the power transfer between transmitting and receiving isotropic radiators is inversely proportional to the distance between them *in wavelengths*. The free space power loss is given by:

$$\text{Free space loss, dB} = 10\log_{10}\frac{(4\pi d)^2}{\lambda^2}$$

where d and λ are in metres, or:

Free space loss (dB) $= 32.4 + 20 \times \log_{10} d + 20 \times \log_{10} f$

where $d =$ distance in km and $f =$ frequency in MHz.

The free space power loss, therefore, increases as the square of the distance and the frequency. Examples are shown in *Figure 1.5*.

With practical antennas, the power gains of the transmitting and receiving antennas, in dBi, must be subtracted from the free space loss calculated as above. Alternatively, the loss may be calculated by:

$$\text{Free space loss (dB)} = 10 \log_{10} \left[\frac{(4\pi d)^2}{\lambda^2} \times \frac{1}{G_t \times G_r} \right]$$

where G_t and G_r are the respective actual gains, not in dB, of the transmitting and receiving antennas.

A major loss in microwave communications and radar systems is atmospheric attenuation (see *Figure 1.6*). The attenuation (in decibels per kilometre (dB/km)) is a function of frequency, with especial problems showing up at 22 GHz and 64 GHz. These spikes are caused by *water vapour* and *atmospheric oxygen* absorption of microwave energy, respectively. Operation of any microwave frequency requires consideration of atmospheric losses, but operation near the two principal spike frequencies poses special problems. At 22 GHz, for example, an additional 1 dB/km of loss must be calculated for the system.

1.5 Behaviour of radio waves

1.5.1 Physical effects

The physical properties of the medium through which a wave travels, and objects in or close to its path, affect the wave in various ways.

Absorption

In the atmosphere absorption occurs and energy is lost in heating the air molecules. Absorption caused by this is minimal at frequencies below about 10 GHz but absorption by foliage, particularly when wet, is severe at VHF and above.

Waves travelling along the earth's surface create currents in the earth causing ground absorption which increases with frequency. A horizontally polarized surface wave suffers more ground absorption than a vertically polarized wave because of the 'short-circuiting' by the ground of the electric field. Attenuation at a given frequency is least for propagation over water and greatest over dry ground for a vertically polarized wave.

Refraction and its effect on the radio horizon

As radio waves travel more slowly in dense media and the densest part of the atmosphere is normally the lowest, the upper parts of a wave usually travel faster than the lower. This refraction (*Figure 1.7*) has the effect of bending the wave to follow the curvature of the earth and progressively tilting the wavefront until eventually the wave becomes horizontally polarized and short-circuited by the earth's conductivity.

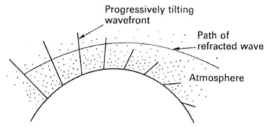

Figure 1.7 Effects of refraction

Waves travelling above the earth's surface (space waves) are usually refracted downwards, effectively increasing the radio horizon to greater than the visual.

The refractive index of the atmosphere is referred to as the K factor; a K factor of 1 indicates zero refraction. Most of the time K is positive at 1.33 and the wave is bent to follow the earth's curvature. The radio horizon is then 4/3 times the visual. However, the density of the atmosphere varies from time to time and in different parts of the world. Density inversions where higher density air is above a region of low density may also occur. Under these conditions the K factor is negative and radio waves are bent away from the earth's surface and are either lost or ducting occurs. A K factor of 0.7 is the worst expected case.

Ducting occurs when a wave becomes trapped between layers of differing density only to be returned at a great distance from its source, possibly creating interference.

Radio horizon distance at VHF/UHF

The radio horizon at VHF/UHF and up is approximately 15% further than the optical horizon. Several equations are used in calculating the distance. If D^- is the distance to the radio horizon, and H is the antenna height, then:

$$D = k\sqrt{H}$$

- When D is in statute miles (5280 feet) and H in feet, then $k = 1.42$.
- When D is in nautical miles (6000 feet) and H in feet, then $k = 1.23$.
- When D is in kilometres and H is in metres, then $k = 4.12$.

Repeating the calculation for the receiving station and adding the results gives the total path length.

Diffraction

When a wave passes over on the edge of an obstacle some of its energy is bent in the direction of the obstacle to provide a signal in what would otherwise be a shadow. The bending is most severe when the wave passes over a sharp edge (*Figure 1.8*).

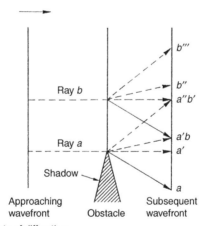

Figure 1.8 Effects of diffraction

As with light waves, the subsequent wavefront consists of wavelets produced from an infinite number of points on the wavefront, rays *a* and *b* in *Figure 1.8* (Huygens' principle). This produces a pattern of interfering waves of alternate addition and subtraction.

Reflection

Radio waves are reflected from surfaces lying in and along their path and also, effectively, from ionized layers in the ionosphere – although most of the reflections from the ionized layers are actually the products of refraction. The strength of truly reflected signals increases with frequency, and the conductivity and smoothness of the reflecting surface.

Multi-path propagation

Reflection, refraction and diffraction may provide signals in what would otherwise be areas of no signal, but they also produce interference.

Reflected – or diffracted – signals may arrive at the receiver in any phase relationship with the direct ray and with each other. The relative phasing of the signals depends on the differing lengths of their paths and the nature of the reflection.

When the direct and reflected rays have followed paths differing by an odd number of half-wavelengths they could be expected to arrive at the receiver in anti-phase with a cancelling effect. However, in the reflection process a further phase change normally takes place. If the reflecting surface had infinite conductivity, no losses would occur in the reflection, and the reflected wave would have exactly the same or opposite phase as the incident wave depending on the polarization in relation to the reflecting surface. In practice, the reflected wave is of smaller amplitude than the incident, and the phase relationships are also changed. The factors affecting the phasing are complex but most frequently, in practical situations, approximately 180° phase change occurs on reflection, so that reflected waves travelling an odd number of half-wavelengths arrive in phase with the direct wave while those travelling an even number arrive anti-phase.

As conditions in the path between transmitter and receiver change so does the strength and path length of reflected signals. This means that a receiver may be subjected to signal variations of almost twice the mean level and practically zero, giving rise to severe fading. This type of fading is frequency selective and occurs on troposcatter systems and in the mobile environment where it is more severe at higher frequencies. A mobile receiver travelling through an urban area can receive rapid signal fluctuations caused by additions and cancellations of the direct and reflected signals at half-wavelength intervals. Fading due to the multi-path environment is often referred to as Rayleigh fading and its effect is shown in *Figure 1.9*. Rayleigh fading, which can cause short signal dropouts, imposes severe restraints on mobile data transmission.

Figure 1.9 Multi-path fading

Noise

The quality of radio signals is not only degraded by the propagation losses: natural or manmade electrical noise is added to them, reducing their intelligibility.

Atmospheric noise includes static from thunderstorms which, unless very close, affects frequencies below about 30 MHz and noise from space is apparent at frequencies between about 8 MHz to 1.5 GHz.

A type of noise with which radio engineers are continually concerned is thermal. Every resistor produces noise spread across the whole frequency spectrum. Its magnitude depends upon the ohmic value of the resistor, its temperature and the bandwidth of the following circuits. The noise voltage produced by a resistor is given by:

$$E_n = \sqrt{4kTBR}$$

where
E_n = noise voltage, V(RMS)
k = Boltzmann's constant
   = 1.38×10^{-23} joules/kelvin
T = temperature in degrees K
B = bandwidth of measurement, Hz
R = resistance in ohms

An antenna possesses resistance and its thermal noise, plus that of a receiver input circuit, is a limiting factor to receiver performance.

Noise is produced in every electronic component. Shot noise – it sounds like falling lead shot – caused by the random arrival of electrons at, say, the collector of a transistor, and the random division of electrons at junctions in devices, add to this noise.

Doppler effect

Doppler effect is an apparent shift of the transmitted frequency which occurs when either the receiver or transmitter is moving. It becomes significant in mobile radio applications towards the higher end of the UHF band and on digitally modulated systems.

When a mobile receiver travels directly towards the transmitter each successive cycle of the wave has less distance to travel before reaching the receiving antenna and, effectively, the received frequency is raised. If the mobile travels away from the transmitter, each successive cycle has a greater distance to travel and the frequency is lowered. The variation in frequency depends on the frequency of the wave, its propagation velocity and the velocity of the vehicle containing the receiver. In the situation where the velocity of the vehicle is small compared with the velocity of light, the frequency shift when moving directly towards, or away from, the transmitter is given to sufficient accuracy for most purposes by:

$$f_d = \frac{V}{C} f_t$$

where
f_d = frequency shift, Hz
f_t = transmitted frequency, Hz
V = velocity of vehicle, m/s
C = velocity of light, m/s

Examples are:

- 100 km/hr at 450 MHz, frequency shift = 41.6 Hz
- 100 km/hr at 1.8 GHz – personal communication network (PCN) frequencies – frequency shift = 166.5 Hz
- Train at 250 km/hr at 900 MHz – a requirement for the GSM pan-European radio-telephone – frequency shift = 208 Hz

When the vehicle is travelling at an angle to the transmitter the frequency shift is reduced. It is calculated as above and the result multiplied by the cosine of the angle of travel from the direct approach (*Figure 1.10*).

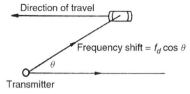

Figure 1.10 Doppler frequency shift and angle to transmitter

In a radar situation Doppler effect occurs on the path to the target and also to the reflected signal so that the above formula is modified to:

$$f_d = \frac{2V}{C} f_t$$

where f_d is now the total frequency shift.

1.6 Methods of propagation

The effects of all of the above phenomena vary with frequency and are used in the selection of frequencies for specific purposes. The behaviour of waves of different frequencies gives rise to the principal types of wave propagation.

Ground wave propagation

Waves in the bands from very low frequencies (VLF, 3–30 kHz), low frequencies (LF, 30–300 kHz) and medium frequencies (MF, 300–3000 kHz) travel close to the earth's surface: the ground wave (*Figure 1.11*). Transmissions using the ground wave must be vertically polarized to avoid the conductivity of the earth short-circuiting the electric field.

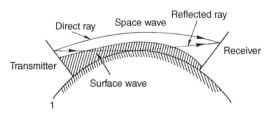

Figure 1.11 Components of the ground wave

The ground wave consists of a surface wave and a space wave. The surface wave travels along the earth's surface, and is attenuated by ground absorption and the tilting of the wavefront due to diffraction.

The losses increase with frequency and thus VLF radio stations have a greater range than MF stations. The attenuation is partially offset by the replacement of energy from the upper part of the wave refracted by the atmosphere.

The calculation of the field strength of the surface wave at a distance from a transmitter is complex and affected by several variables. Under plane earth conditions and when the distance is sufficiently short that the earth's curvature can be neglected the field intensity is given by:

$$E_{su} = \frac{2E_0}{d} A$$

where

E_{su} = field intensity in same units as E_0

d = distance in same units of distance as used in E_0

A = a factor calculated from the earth losses, taking frequency, dielectric constant and conductivity into account

E_0 = the free space field produced at unit distance from the transmitter. (With a short (compared with $\lambda/4$) vertical aerial, $2E_0 = 300\sqrt{P}$ mV/m at 1 km where P is the radiated power in kW.) (Terman, 1943)

For a radiated power of 1 kW and ground of average dampness, the distance at which a field of 1 mV/m will exist is given in *Table 1.2*.

Table 1.2 Distance at which a field of 1 mV/m will exist for a radiated power of 1 kW and ground of average dampness

Frequency	Range (km)
100 kHz	200
1 MHz	60
10 MHz	6
100 MHz	1.5

The direct and reflected components of the ground wave produce multi-path propagation and variations in received single strength will arise depending on the different path lengths taken by the two components. When the transmitting and receiving antennas are at ground level the components of the space wave cancel each other and the surface wave is dominant. When the antennas are elevated, the space wave becomes increasingly strong and a height is eventually reached where the surface wave has a negligible effect on the received signal strength.

Sky wave propagation

High frequency (HF) waves between 3 MHz and 30 MHz are effectively reflected by ionized layers in the ionosphere producing the sky wave. Medium frequency waves may also be reflected, but less reliably.

The ionosphere contains several layers of ionized air at varying altitudes (*Figure 1.12*). The heights and density of the layers vary diurnally, seasonally and with the incidence of sunspot activity. The E and F_2 layers are semi-permanent while the F_1 layer is normally only present during daytime.

Radio waves radiated at a high angle and reflected by these layers return to earth at a distance from the transmitter. The HF reflection

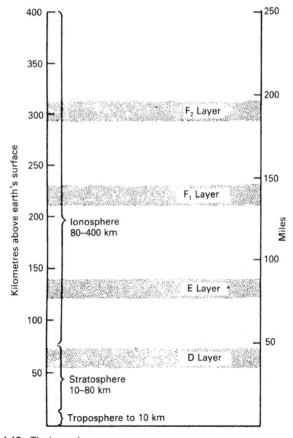

Figure 1.12 The ionosphere

process is in reality one of refraction in layers possessing a greater free electron density than at heights above or below them. The speed of propagation is slowed on entering a layer and the wave is bent and, if of a suitable frequency and angle of incidence, returned to earth (*Figure 1.13*). The terms used are defined as follows:

- *Virtual height.* The height at which a true reflection of the incident wave would have occurred (*Figure 1.13*).
- *Critical frequency* (f_c). The highest frequency that would be returned to earth in a wave directed vertically at the layer.
- *Maximum usable frequency* (muf). The highest frequency that will be returned to earth for a given angle of incidence. If the angle of incidence to the normal is θ, the muf $= f_c / \cos \theta$.
- *Skip distance.* The minimum distance from the transmitter, along the surface of the earth, at which a wave above the critical frequency will be returned to earth (*Figure 1.12*). Depending on the frequency, the ground wave will exist at a short distance from the transmitter.
- *Sporadic E-layer reflections.* Reflections from the E layer at frequencies higher than those which would normally be returned to earth. They appear to be reflections from electron clouds having sharp boundaries and drifting through the layer. As the name implies the reflections are irregular but occur mostly in summer and at night.

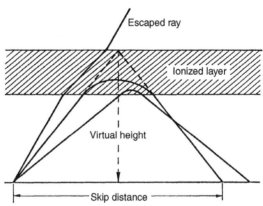

Figure 1.13 Sky wave propagation

Space wave propagation

The space wave travels through the troposphere (the atmosphere below the ionosphere) between the transmitter and the receiver. It contains

Figure 1.14 Pictorial representation of radio coverage from a base station

Refracted wave

Diffracted wave

Area of signal from reflected wave

Area of shadow

both direct and reflected components (see *Figure 1.11*), and is affected by refraction and diffraction. The importance of these effects varies with frequency, the nature of the terrain and of objects close to the direct path, and the type of communication, e.g. data. Apart from medium-wave broadcasting, space waves are used mainly for communications using frequencies of VHF and upwards.

The range of space waves is the radio horizon. However, places of little or no signal can arise in the lee of radio obstacles. Fortunately, they may be filled with either reflected or diffracted signals as depicted in *Figure 1.14*.

Tropospheric scatter

The tropospheric, or forward, scatter effect provides reliable, over the horizon, communication between fixed points at bands of ultra and super high frequencies. Usable bands are around 900, 2000 and 5000 MHz and path lengths of 300 to 500 km are typical.

The mechanism is not known with certainty but reflections from discontinuities in the dielectric constant of the atmosphere and scattering of the wave by clouds of particles are possibilities. It is an inefficient process, the scattered power being −60 to −90 dB relative to the incident power, so high transmitter powers are necessary. The phenomenon is regularly present but is subject to two types of fading. One occurs slowly and is due to variations of atmospheric conditions. The other is a form of Rayleigh fading and is rapid, deep and frequency selective. It is due to the scattering occurring at different points over a volume in the atmosphere producing multipath propagation conditions.

Troposcatter technique uses directional transmitting and receiving antennas aimed so that their beams intercept in the troposphere at the mid-distance point. To overcome the fading, diversity reception using multiple antennas spaced over 30 wavelengths apart is common.

1.7 Other propagation topics

Communications in the VHF through microwave regions normally takes place on a 'line-of-sight' basis where the radio horizon defines the limit of sight. In practice, however, the situation is not so neat and simple. There is a transition region between the HF and VHF where long distance ionospheric 'skip' occurs only occasionally. This effect is seen above 25 MHz, and is quite pronounced in the 50 MHz region. Sometimes the region behaves like line-of-sight VHF, and at others like HF shortwave.

1.7.1 Scatter

There are a number of scatter modes of propagation. These modes can extend the radio horizon a considerable amount. Where the radio horizon might be a few tens of kilometres, underscatter modes permit very much longer propagation. For example, a local FM broadcaster at 100 MHz might have a service area of about 40 miles, and might be heard 180 miles away during the summer months when *Sporadic-E* propagation occurs. One summer, a television station in Halifax, Nova Scotia, Canada, was routinely viewable in Washington, DC in the United States during the early morning hours for nearly a week.

Sporadic-E is believed to occur when a small region of the atmosphere becomes differentially ionized, and thereby becomes a species of 'radio mirror'. Ionospheric scatter propagation occurs when clouds of ions exist in the atmosphere. These clouds can exist in both the ionosphere and the troposphere, although the tropospheric model is more reliable for communications. A signal impinging this region may be scattered towards other terrestrial sites which may be a great distance away. The specific distance depends on the geometry of the scenario.

There are at least three different modes of scatter from ionized clouds: *back scatter, side scatter*, and *forward scatter*. The back scatter mode is a bit like radar, in that signal is returned back to the transmitter site, or in regions close to the transmitter. Forward scatter occurs when the reflected signal continues in the same azimuthal direction (with respect to the transmitter), but is redirected toward the Earth's surface. Side scatter is similar to forward scatter, but the azimuthal direction might change.

Unfortunately, there are often multiple reflections from the ionized cloud, and these are shown as 'multiple scatter' in *Figure 1.15*. When these reflections are able to reach the receiving site, the result is a rapid, fluttery fading that can be of quite profound depths.

Meteor scatter is used for communication in high latitude regions. When a meteor enters the Earth's atmosphere it leaves an ionized trail of air behind it. This trail might be thousands of kilometres long, but is very short lived. Radio signals impinging the tubular metre ion trail are reflected back towards Earth. If the density of meteors in the critical region is high, then more or less continuous communications can be achieved. This phenomenon is noted in the low VHF between 50 and about 150 MHz. It can easily be observed on the FM broadcast band if the receiver is tuned to distant stations that are barely audible. If the geometry of the scenario is right, abrupt but short-lived peaks in the signal strength will be noted.

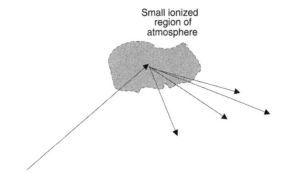

Figure 1.15 Multiple scatter

1.7.2 Refraction modes

Refraction is the mechanism for most tropospheric propagation phenomena. The dielectric properties of the air, which are set mostly by the moisture content, are a primary factor in tropospheric refraction. Refraction occurs in both light or radio wave systems when the wave passes between mediums of differing density. Under that situation, the wave path will bend an amount proportional to the difference in density of the two regions.

The general situation is typically found at UHF and microwave frequencies. Because air density normally decreases with altitude, the top of a beam of radio waves typically travels slightly faster than the lower portion of the beam. As a result, those signals refract a small amount. Such propagation provides slightly longer surface distances than are normally expected from calculating the distance to the radio horizon. This phenomenon is called *simple refraction*, and is described by the *K factor*.

Super refraction

A special case of refraction called super refraction occurs in areas of the world where warmed land air flows out over a cooler sea (*Figure 1.16*). Examples of such areas are deserts that are adjacent to a large body of water: the Gulf of Aden, the southern Mediterranean, and the Pacific Ocean off the coast of Baja, California. Frequent VHF/UHF/microwave communications up to 200 miles are reported in such areas, and up to 600 miles have reportedly been observed.

Ducting

Another form of refraction phenomenon is weather related. Called *ducting*, this form of propagation is actually a special case of super

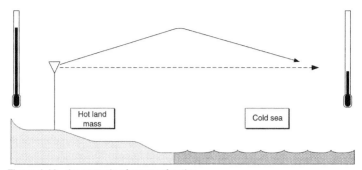

Figure 1.16 An example of super refraction

refraction. Evaporation of sea water causes temperature inversion regions to form in the atmosphere. That is, layered air masses in which the air temperature is greater than in the layers below it (note: air temperature normally decreases with altitude, but at the boundary with an inversion region, it begins to increase). The inversion layer forms a 'duct' that acts similarly to a waveguide. Ducting allows long distance communications from lower VHF through microwave frequencies; with 50 MHz being a lower practical limit, and 10 GHz being an ill-defined upper limit. Airborne operators of radio, radar, and other electronic equipment can sometimes note ducting at even higher microwave frequencies.

Antenna placement is critical for ducting propagation. Both the receiving and transmitting antennas must be either: (a) physically inside the duct (as in airborne cases), or (b) able to propagate at an angle such that the signal gets trapped inside the duct. The latter is a function of antenna radiation angle. Distances up to 2500 miles or so are possible through ducting.

Certain paths, where frequent ducting occurs, have been identified: in the United States, the Great Lakes region to the southeastern Atlantic seaboard; Newfoundland to the Canary Islands; across the Gulf of Mexico from Florida to Texas; Newfoundland to the Carolinas; California to Hawaii; and Ascension Island to Brazil.

Subrefraction

Another refractive condition is noted in the polar regions, where colder air from the land mass flows out over warmer seas (*Figure 1.17*). Called *subrefraction*, this phenomena bends EM waves away from the Earth's surface – thereby reducing the radio horizon by about 30 to 40%.

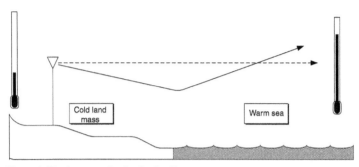

Figure 1.17 An example of subrefraction

All tropospheric propagation that depends upon air-mass temperatures and humidity shows diurnal (i.e. over the course of the day) variation caused by the local rising and setting of the sun. Distant signals may vary 20 dB in strength over a 24-hour period. These tropospheric phenomena explain how TV, FM broadcast, and other VHF signals can propagate great distances, especially along seacoast paths, sometimes weak and sometimes nonexistent.

1.7.3 Great circle paths

A great circle is the shortest line between two points on the surface of a sphere, such that it lays on a plane through the Earth's centre and includes the two points. When translated to 'radiospeak', a great circle is the shortest path on the surface of the Earth between two points. Navigators and radio operators use the great circle for similar, but different, reasons. Navigators use it in order to get from here to there, and radio operators use it to get a transmission path from here to there.

The heading of a directional antenna is normally aimed at the receiving station along its great circle path. Unfortunately, many people do not understand the concept well enough, for they typically aim the antenna in the wrong direction. For example, Washington, DC in the USA is on approximately the same latitude as Lisbon, Portugal. If you fly due east, you will have dinner in Lisbon, right? Wrong. If you head due east from Washington, DC, across the Atlantic, the first landfall would be West Africa, somewhere between Ghana and Angola. Why? Because the great circle bearing 90 degrees takes us far south. The geometry of spheres, not flat planes, governs the case.

Long path versus short path

The Earth is a sphere (or, more precisely, an 'oblique spheroid'), so from any given point to any other point there are two great circle paths: the long path (major arc) and the short path (minor arc). In general, the best reception occurs along the short path. In addition, short path propagation is more nearly 'textbook', compared with long path reception. However, there are times when long path is better, or is the only path that will deliver a signal to a specific location from the geographic location in question.

Grey line propagation

The Grey line is the twilight zone between the night and daytime halves of the earth. This zone is also called the *planetary terminator*. It varies up to +23 degrees either side of the north–south longitudinal lines, depending on the season of the year (it runs directly north–south only at the vernal and autumnal equinoxes). The D-layer of the iono-sphere absorbs signals in the HF region. This layer disappears almost completely at night, but it builds up during the day. Along the grey line, the D-layer is rapidly decaying west of the line, and has not quite built up east of the line.

Brief periods of abnormal propagation occur along the grey line. Stations on either side of the line can be heard from regions, and at distances, that would otherwise be impossible on any given frequency. For this reason, radio operators often prefer to listen at dawn and dusk for this effect.

1.7.4 Scatter propagation modes

Auroral propagation

The auroral effect produces a luminescence in the upper atmosphere resulting from bursts of particles released from the sun 18 to 48 hours earlier. The light emitted is called the northern lights and the southern lights. The ionized regions of the atmosphere that create the lights form a radio reflection shield, especially at VHF and above, although 15 to 20 MHz effects are known. Auroral propagation effects are normally seen in the higher latitudes, although listeners in the southern tier of states in the USA and Europe are often treated to the reception of signals from the north being reflected from auroral clouds. Similar effects exist in the southern hemisphere.

Non-reciprocal direction

If you listen to the 40 metre (7–7.3 MHz) amateur radio band receiver on the East Coast of the United States, you will sometimes hear European stations – especially in the late afternoon. But when the

US amateur tries to work those European stations there is no reply whatsoever. The Europeans are unable to hear the US stations. This propagation anomaly causes the radio wave to travel different paths dependent on which direction it travels, i.e. an east–west signal is not necessarily the reciprocal of a west–east signal. This anomaly can occur when a radio signal travels through a heavily ionized medium in the presence of a magnetic field, which is exactly the situation when the signal travels through the ionosphere in the presence of the Earth's magnetic field.

References

Belcher, R. *et al.* (1989). *Newnes Mobile Radio Servicing Handbook.* Butterworth-Heinemann, Oxford.

Kennedy, G. (1977). *Electronic Communications Systems.* McGraw-Hill Kogashuka, Tokyo.

Terman, F.E. (1943). *Radio Engineers' Handbook.* McGraw-Hill, London.

2 The decibel scale

2.1 Decibels and the logarithmic scale

The range of powers, voltages and currents encountered in radio engineering is too wide to be expressed on linear scale. Consequently, a logarithmic scale based on the decibel (dB, one tenth of a bel) is used. The decibel does not specify a magnitude of a power, voltage or current but a ratio between two values of them. Gains and losses in circuits or radio paths are expressed in decibels.

The ratio between two powers is given by:

$$\text{Gain or loss, dB} = 10 \log_{10} \frac{P_1}{P_2}$$

where P_1 and P_2 are the two powers.

As the power in a circuit varies with the square of the voltage or current, the logarithm of the ratio of these quantities must be multiplied by twenty instead of ten. To be accurate the two quantities under comparison must operate in identical impedances:

$$\text{Gain or loss, dB} = 20 \log_{10} \frac{V_1}{V_2}$$

To avoid misunderstandings, it must be realized that a ratio of 6 dB is 6 dB regardless of whether it is power, voltage or current that is referred to: if it is power, the ratio for 6 dB is four times; if it is voltage or current, the ratio is two times (*Table 2.1*).

2.2 Decibels referred to absolute values

While the decibel scale expresses ratios only, if a reference value is added to the statement as a suffix it can be used to refer to absolute values. For example, a loss of 10 dB means a reduction in power to a level equal to one tenth of the original and if the statement is −10 dBm the level referred to is 1/10 of a milliwatt. Commonly used suffixes and, where applicable, their absolute reference levels are as follows. *Table 2.2* shows the relative levels in decibels at 50 ohms impedance.

Table 2.1 The decibel figures are in the centre column: figures to the left represent decibel loss, and those to the right decibel gain. The voltage and current figures are given on the assumption that there is no change in impedance

Voltage or current ratio	Power ratio	dB ← − + →	Voltage or current ratio	Power ratio
1.000	1.000	0	1.000	1.000
0.989	0.977	0.1	1.012	1.023
0.977	0.955	0.2	1.023	1.047
0.966	0.933	0.3	1.035	1.072
0.955	0.912	0.4	1.047	1.096
0.944	0.891	0.5	1.059	1.122
0.933	0.871	0.6	1.072	1.148
0.912	0.832	0.8	1.096	1.202
0.891	0.794	1.0	1.122	1.259
0.841	0.708	1.5	1.189	1.413
0.794	0.631	2.0	1.259	1.585
0.750	0.562	2.5	1.334	1.778
0.708	0.501	3.0	1.413	1.995
0.668	0.447	3.5	1.496	2.239
0.631	0.398	4.0	1.585	2.512
0.596	0.355	4.5	1.679	2.818
0.562	0.316	5.0	1.778	3.162
0.501	0.251	6.0	1.995	3.981
0.447	0.200	7.0	2.239	5.012
0.398	0.159	8.0	2.512	6.310
0.355	0.126	9.0	2.818	7.943
0.316	0.100	10	3.162	10.00
0.282	0.0794	11	3.55	12.6
0.251	0.0631	12	3.98	15.9
0.224	0.0501	13	4.47	20.0
0.200	0.0398	14	5.01	25.1
0.178	0.0316	15	5.62	31.6
0.159	0.0251	16	6.31	39.8
0.126	0.0159	18	7.94	63.1
0.100	0.0100	20	10.00	100.0
3.16×10^{-2}	10^{-3}	30	3.16×10	10^3
10^{-2}	10^{-4}	40	10^2	10^4
3.16×10^{-3}	10^{-5}	50	3.16×10^2	10^5
10^{-3}	10^{-6}	60	10^3	10^6
3.16×10^{-4}	10^{-7}	70	3.16×10^3	10^7
10^{-4}	10^{-8}	80	10^4	10^8
3.16×10^{-5}	10^{-9}	90	3.16×10^4	10^9
10^{-5}	10^{-10}	100	10^5	10^{10}
3.16×10^{-6}	10^{-11}	110	3.16×10^5	10^{11}
10^{-6}	10^{-12}	120	10^6	10^{12}

Table 2.2 Relative levels in decibels at 50 ohms impedance

dB μV	Voltage	dBV	dBm	dBW	Power
−20	100 nV	−140	−127	−157	200 aW
−19	115	−139	−126	−156	250
−18	125	−138	−125	−155	315
−17	140	−137	−124	−154	400
−16	160	−136	−123	−153	500
−15	180	−135	−122	−152	630
−14	200	−134	−121	−151	800
−13	225	−133	−120	−150	1 fW
−12	250	−132	−119	−149	1.25
−11	280	−131	−118	−148	1.6
−10	315	−130	−117	−147	2.0
−9	355	−129	−116	−146	2.5
−8	400	−128	−115	−145	3.15
−7	450	−127	−114	−144	4.0
−6	500	−126	−113	−143	5.0
−5	565	−125	−112	−142	6.3
−4	630	−124	−111	−141	8.0
−3	710	−123	−110	−140	10
−2	800	−122	−109	−139	12.5
−1	900	−121	−108	−138	16
0	1 μV	−120	−107	−137	20
1	1.15	−119	−106	−136	25
2	1.25	−118	−105	−135	31.5
3	1.4	−117	−104	−134	40
4	1.6	−116	−103	−133	50
5	1.8	−115	−102	−132	63
6	2.0	−114	−101	−131	80
7	2.25	−113	−100	−130	100
8	2.5	−112	−99	−129	125
9	2.8	−111	−98	−128	160
10	3.15	−110	−97	−127	200
11	3.55	−109	−96	−126	250
12	4.0	−108	−95	−125	315
13	4.5	−107	−94	−124	400
14	5.0	−106	−93	−123	500
15	5.65	−105	−92	−122	630
16	6.3	−104	−91	−121	800
17	7.1	−103	−90	−120	1 pW
18	8.0	−102	−89	−119	1.25
19	9.0	−101	−88	−118	1.6
20	10.0	−100	−87	−117	2.0
21	11.5	−99	−86	−116	2.5
22	12.5	−98	−85	−115	3.15
23	14	−97	−84	−114	4.0
24	16	−96	−83	−113	5.0
25	18	−95	−82	−112	6.3

(*continued overleaf*)

Table 2.2 (continued)

dB μV	Voltage	dBV	dBm	dBW	Power
26	20	−94	−81	−111	8.0
27	22.5	−93	−80	−110	10
28	25	−92	−79	−109	12.5
29	28	−91	−78	−108	16
30	31.5	−90	−77	−107	20
31	35.5	−89	−76	−106	25
32	40	−88	−75	−105	31.5
33	45	−87	−74	−104	40
34	50	−86	−73	−103	50
35	56.5	−85	−72	−102	63
36	63	−84	−71	−101	80
37	71	−83	−70	−100	100
38	80	−82	−69	−99	125
39	90	−81	−68	−98	160
40	100	−80	−67	−97	200
41	115	−79	−66	−96	250
42	125	−78	−65	−95	315
43	140	−77	−64	−94	400
44	160	−76	−63	−93	500
45	180	−75	−62	−92	630
46	200	−74	−61	−91	800
47	225	−73	−60	−90	1 nW
48	250	−72	−59	−89	1.25
49	280	−71	−58	−88	1.6
50	315	−70	−57	−87	2.0
51	355	−69	−56	−86	2.5
52	400	−68	−55	−85	3.15
53	450	−67	−54	−84	4.0
54	500	−66	−53	−83	5.0
55	565	−65	−52	−82	6.3
56	630	−64	−51	−81	8.0
57	710	−63	−50	−80	10
58	800	−62	−49	−79	12.5
59	900	−61	−48	−78	16
60	1 mV	−60	−47	−77	20
61	1.15	−59	−46	−76	25
62	1.25	−58	−45	−75	31.5
63	1.4	−57	−44	−74	40
64	1.6	−56	−43	−73	50
65	1.8	−55	−42	−72	63
66	2.0	−54	−41	−71	80
67	2.25	−53	−40	−70	100
68	2.5	−52	−39	−69	125
69	2.8	−51	−38	−68	160
70	3.15	−50	−37	−67	200
71	3.55	−49	−36	−66	250
72	4.0	−48	−35	−65	315

Table 2.2 (continued)

dB μV	Voltage	dBV	dBm	dBW	Power
73	4.5	−47	−34	−64	400
74	5.0	−46	−33	−63	500
75	5.65	−45	−32	−62	630
76	6.3	−44	−31	−61	800
77	7.1	−43	−30	−60	1 μW
78	8.0	−42	−29	−59	1.25
79	9.0	−41	−28	−58	1.6
80	10 mV	−40	−27	−57	2.0
81	11.5	−39	−26	−56	2.5
82	12.5	−38	−25	−55	3.15
83	14	−37	−24	−54	4.0
84	16	−36	−23	−53	5.0
85	18	−35	−22	−52	6.3
86	20	−34	−21	−51	8.0
87	22.5	−33	−20	−50	10
88	25	−32	−19	−49	12.5
89	28	−31	−18	−48	16
90	31.5	−30	−17	−47	20
91	35.5	−29	−16	−46	25
92	40	−28	−15	−45	31.5
93	45	−27	−14	−44	40
94	50	−26	−13	−43	50
95	56.5	−25	−12	−42	63
96	63	−24	−11	−41	80
97	71	−23	−10	−40	100
98	80	−22	−9	−39	125
99	90	−21	−8	−38	160
100	100	−20	−7	−37	200
101	115	−19	−6	−36	250
102	125	−18	−5	−35	315
103	140	−17	−4	−34	400
104	160	−16	−3	−33	500
105	180	−15	−2	−32	630
106	200	−14	−1	−31	800
107	225	−13	0	−30	1 mW
108	250	−12	1	−29	1.25
109	280	−11	2	−28	1.6
110	315	−10	3	−27	2.0
111	355	−9	4	−26	2.5
112	400	−8	5	−25	3.15
113	450	−7	6	−24	4.0
114	500	−6	7	−23	5.0
115	565	−5	8	−22	6.3
116	630	−4	9	−21	8.0
117	710	−3	10	−20	10
118	800	−2	11	−19	12.5

(continued overleaf)

Table 2.2 (*continued*)

dB μV	Voltage	dBV	dBm	dBW	Power
119	900	−1	12	−18	16
120	1 V	0	13	−17	20
121	1.15	1	14	−16	25
122	1.25	2	15	−15	31.5
123	1.4	3	16	−14	40
124	1.6	4	17	−13	50
125	1.8	5	18	−12	63
126	2.0	6	19	−11	80
127	2.25	7	20	−10	100
128	2.5	8	21	−9	125
129	2.8	9	22	−8	160
130	3.15	10	23	−7	200
131	3.55	11	24	−6	250
132	4.0	12	25	−5	315
133	4.5	13	26	−4	400
134	5.0	14	27	−3	500
135	5.65	15	28	−2	630
136	6.3	16	29	−1	800
137	7.1	17	30	0	1 W
138	8.0	18	31	1	1.25
139	9.0	19	32	2	1.6
140	10	20	33	3	2.0
141	11.5	21	34	4	2.5
142	12.5	22	35	5	3.15
143	14	23	36	6	4.0
144	16	24	37	7	5.0
145	18	25	38	8	6.3
146	20	26	39	9	8.0
147	22.5	27	40	10	10
148	25	28	41	11	12.5
149	28	29	42	12	16
150	31.5	30	43	13	20
151	35.5	31	44	14	25
152	40	32	45	15	31.5
153	45	33	46	16	40
154	50	34	47	17	50
155	56.5	35	48	18	63
156	63	36	49	19	80
157	71	37	50	20	100
158	80	38	51	21	125
159	90	39	52	22	160
160	100	40	53	23	200
161	115	41	54	24	250
162	125	42	55	25	315
163	140	43	56	26	400
164	160	44	57	27	500
165	180	45	58	28	630

Table 2.2 (continued)

dB μV	Voltage	dBV	dBm	dBW	Power
166	200	46	59	29	800
167	225	47	60	30	1 kW
168	250	48	61	31	1.25
169	280	49	62	32	1.6
170	315	50	63	33	2.0
171	355	51	64	34	2.5
172	400	52	65	35	3.15
173	450	53	66	36	4.0
174	500	54	67	37	5.0
175	565	55	68	38	6.3
176	630	56	69	39	8.0
177	710	57	70	40	10
178	800	58	71	41	12.5
179	900	59	72	42	16
180	1 kV	60	73	43	20

Decibel glossary

dBa   stands for dB 'adjusted'. This is a weighted circuit noise power referred to −85 dBm, which is 0 dBa. (Historically measured with a noise meter at the receiving end of a line. The meter is calibrated on a 1000 Hz tone such that 1 mW (0 dBm) gives a reading of +85 dBm. If the 1 mW is spread over the band 300–3400 Hz as random white noise, the meter will read +82 dBa.)

dBa0   circuit noise power in dBa referred to, or measured at, a point of zero relative transmission level (0 dBr). (A point of zero relative transmission level is a point arbitrarily established in a transmission circuit. All other levels are stated relative to this point.) It is preferable to convert circuit noise measurement values from dBa to dBa0 as this makes it unnecessary to know or to state the relative transmission level at the point of measurement.

dBd   used for expressing the gain of an antenna referred to a dipole.

dBi   used for expressing the gain of an antenna referred to an isotropic radiator.

dB μV   decibels relative to 1 microvolt.

dbm   decibels relative to 1 milliwatt. The term dBm was originally used for telephone and audio work and, when used in that context, implies an impedance of 600 Ω, the nominal impedance of a telephone line. When it is desired to define a relative transmission level in a circuit, dBr is preferred.

Table 2.3 Binary decibel values

Bits	Max. value	Decibels (dB)
1	2	6.02
2	4	12.04
3	8	18.06
4	16	24.08
5	32	30.10
6	64	36.12
7	128	42.14
8	256	48.16
9	512	54.19
10	1024	60.21
11	2048	66.23
12	4096	72.25
13	8192	78.27
14	16 384	84.29
15	32 768	90.31
16	65 536	96.33
17	131 072	102.35
18	262 144	108.37
19	524 288	114.39
20	1 048 576	120.41
21	2 097 152	126.43
22	4 194 304	132.45
23	8 388 608	138.47
24	16 777 216	144.49
25	33 554 432	150.51
26	67 108 864	156.54
27	134 217 728	162.56
28	268 435 456	168.58
29	536 870 912	174.60
30	1 073 741 824	180.62
31	2 147 483 648	186.64
32	4 294 967 296	192.66

dBm0   dBm referred to, or measured at, a point of zero transmission level.

dBmp   a unit of noise power in dBm, measured with psophometric weighting. $dBmp = 10\log_{10} pWp - 90 = dBa - 84 = dBm - 2.5$ (for flat noise 300–3400 Hz).

         pWp = picowatts psophometrically weighted.

dBm0p   the abbreviation for absolute noise power in dBm referred to or measured at a point of zero relative transmission level, psophometrically weighted.

dBr   means dB 'relative level'. Used to define transmission level at various points in a circuit or system referenced to the zero transmission level point.

dBrn   a weighted circuit noise power unit in dB referenced to 1 pW
        (-90 dBm) which is 0 dBrn.
dBrnc   weighted noise power in dBrn, measured by a noise measuring
        set with 'C-message' weighting.
dBrnc0 noise measured in dBrnc referred to zero transmission level
        point.
dBu   decibels relative to 0.775 V, the voltage developed by 1 mW
        when applied to 600 Ω. dBu is used in audio work when the
        impedance is not 600 Ω and no specific impedance is implied.
dbV   decibels relative to 1 volt.
dbW   decibels relative to 1 watt.

Note: To convert dBm to dB μV add 107 (e.g. -20 dBm $= -20 +$
$107 = 87$ dB μV.

The beauty of decibel notation is that system gains and losses can
be computed using addition and subtraction rather than multiplication
and division. For example, suppose a system consists of an antenna
that delivers a -4.7 dBm signal at its terminals (we conveniently
neglect the antenna gain by this ploy). The antenna is connected to
a 40 dB low-noise amplifier (A1) at the head end, and then through
a 370 metre long coaxial cable to a 20 dB gain amplifier (A2), with
a loss (L1) of -48 dB. The amplifier is followed by a bandpass filter with a -2.8 dB insertion loss (L2), and a -10 dB attenuator (L3).
How does the signal exist at the end of this cascade chain?

S1	-4.7 dBm
A1	40.0 dB
A2	20.0 dB
L1	-48.0 dB
L2	-2.8 dB
L3	-10.0 dB
Total:	-5.5 dBm

Converting dBm to watts

$$P = \frac{10^{\text{dBm}/10}}{1000}$$

Converting any dB to ratio

Power levels: $\dfrac{P1}{P2} = 10^{\text{dB}/10}$

Voltage levels: $\dfrac{V1}{V2} = 10^{dB/20}$

Current levels: $\dfrac{I1}{I2} = 10^{dB/20}$

Binary decibel values

Binary numbers are used in computer systems. With the digitization of RF systems it is necessary to understand the decibel values of binary numbers. These binary numbers might be from an analogue-to-digital converter (ADC or A/D) that digitizes the IF amplifier output, or a digital-to-analogue converter (DAC or D/A) used to generate the analogue signal in a direct digital synthesis (DDS) signal generator.

3 Transmission lines

3.1 General considerations

The purpose of any transmission line is to transfer power between a source and a load with the minimum of loss and distortion in either amplitude, frequency or phase angle.

Electrons travel more slowly in conductors than they do in free space and all transmission lines contain distributed components: resistance, inductance and capacitance. Consequently, lines possess an impedance which varies with frequency, and loss and distortion occur. Because the impedance is not constant over a wide frequency band the insertion loss will not be the same for all frequencies and frequency distortion will arise. A wavefront entering a line from a source takes a finite time to travel its length. This transit time, because of the distributed components, also varies with frequency and creates phase distortion.

3.2 Impedance matching

To transfer the maximum power from a generator into a load the impedance of the load and the internal impedance of the generator – and any intervening transmission line – must be equal.

Figure 3.1 illustrates the simplest case of a generator of internal impedance Z_s equal to 5 ohms and producing an e.m.f. of 20 volts.

When loads of varying impedance, Z_1, are connected the output voltage, V (p.d.) and the power in the load, P_1, varies as follows:

$Z_1 = 5\,\Omega$ $I = 20/10 = 2\,\text{A}$ $V = 10\,\text{V}$
$Z_1 = 3\,\Omega$ $I = 2.5\,\text{A}$ $V = 7.5\,\text{V}$
$Z_1 = 8.33\,\Omega$ $I = 1.5\,\text{A}$ $V = 12.5\,\text{V}$

$P_1 = V^2/Z_1 = 100/5 = 20\,\text{W}$
$P_1 = V^2/Z_1 = 56.25/3 = 18.75\,\text{W}$
$P_1 = V^2/Z_1 = 156.25/8.33 = 18.75\,\text{W}$

When dealing with alternating current and when transmission lines, particularly at radio frequencies, are interposed between the source and the load, other factors than the power transfer efficiency must also be considered.

Figure 3.1 Impedance matching

3.3 Base band lines

These are the lines which generally operate at comparatively low frequencies carrying information at base band, e.g. speech, music, video or data. Generally provided by the telecommunications or telephone companies, usually on a rental basis, they are no longer likely to be hard wired, solid copper lines, although these may still be obtainable for lengths below about 10 km within one exchange area. Longer lines will probably be multiplexed, and comprised of radio and optical circuits over part of their length.

Baseband line impedance may vary between 450 Ω and 750 Ω. Nominal impedance is 600 Ω. Most line parameters are specified when measured between 600 Ω non-reactive impedance.

3.4 Balanced line hybrids

Radio transmitters and receivers are often controlled over a two-wire line. To facilitate this a balanced line hybrid circuit, consisting of two transformers connected back to back as in *Figure 3.2*, is inserted between the transmitter and receiver, and the line.

A signal from the receiver audio output is fed to winding L_1 of transformer T_1 which induces voltages across L_2 and L_3. The resultant line current also flows through L_4 and produces a voltage across L_6 which would appear as modulation on the transmitter but for the anti-phase voltage appearing across L_5. To ensure that the voltages cancel exactly a variable resistor, and often a capacitor to equalize the frequency response, is connected between L_2 and L_5.

A signal arriving via the line is applied to the transmitter as modulation; that it is also applied to the receiver poses no problem.

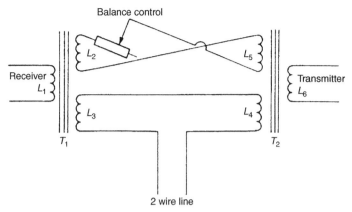

Figure 3.2 Balanced line hybrid

3.5 Radio frequency lines

Radio frequency transmission lines possess similar electrical characteristics to base band lines. However, they may be carrying large powers and the effects of a mismatched load are much more serious than a loss of transferred power. Three types of wire RF line are commonly used: a single wire with ground return for MF and LF broadcast transmission, an open pair of wires at HF and co-axial cable at higher frequencies. Waveguides are used at the higher microwave frequencies. RF lines exhibit an impedance characterized by their type and construction.

3.5.1 Characteristic impedance, Z_0

The physical dimensions of an RF transmission line, the spacing between the conductors, their diameters and the dielectric material between them, determine the characteristic impedance of the line, Z_0, which is calculated for the most commonly used types as follows.

Single wire with ground return (*Figure 3.3(a)*):

$$Z_0 = 138 \log_{10} \frac{4h}{d} \text{ ohms}$$

2-wire balanced, in air (*Figure 3.3(b)*):

$$Z_0 = \frac{276}{\sqrt{\varepsilon_p}} \log_{10} \frac{2s}{d} \text{ ohms}$$

(a) Single wire, ground return line

(b) 2-wire balanced air dielectric line

(c) Co-axial cable

Figure 3.3

Co-axial (*Figure 3.3(c)*):

$$Z_0 = \frac{138}{\sqrt{\varepsilon_r}} \log_{10} \frac{R}{r} \text{ ohms}$$

where dimensions are in mm and ε_r = relative permittivity of continuous dielectric.

3.5.2 Insertion loss

The loss in RF cables is quoted in specifications as attenuation in dB per unit length at specific frequencies, the attenuation increasing with frequency. The electrical specifications for cables having a braided outer conductor are given in *Tables 3.1* and *3.2*. Those for the now commonly used foam dielectric, solid outer conductor cables are provided in *Table 3.3*.

Table 3.1 RF cables USA RG and URM series

RG number	Nominal impedance Z0 (ohms)	Overall diameter (inches)	Velocity factor	Attenuation (dB per 100 ft)					Capacity (pF/ft)	Maximum operating voltage RMS
				1 MHz	10 MHz	100 MHz	1000 MHz	3000 MHz		
RG-6A/U	75.0	0.332	0.659	0.21	0.78	2.9	11.2	21.0	20.0	2700
RG-11A/U	75.0	0.405	0.66	0.18	0.7	2.3	7.8	16.5	20.5	5000
RG-12A/U	75.0	0.475	0.659	0.18	0.66	2.3	8.0	16.5	20.5	4000
RG-16/U	52.0	0.630	0.670	0.1	0.4	1.2	6.7	16.0	29.5	6000
RG-34A/U	75.0	0.630	0.659	0.065	0.29	1.3	6.0	12.5	20.5	5200
RG-34B/U	75.0	0.630	0.66	–	0.3	1.4	5.8	–	21.5	6500
RG-35A/U	75.0	0.945	0.659	0.07	0.235	0.85	3.5	8.60	20.5	10 000
RG-54A/U	58.0	0.250	0.659	0.18	0.74	3.1	11.5	21.5	26.5	3000
RG-55/U	53.5	0.206	0.659	0.36	1.3	4.8	17.0	32.0	28.5	1900
RG-55A/U	50.0	0.216	0.659	0.36	1.3	4.8	17.0	32.0	29.5	1900
RG-58/U	53.5	0.195	0.659	0.33	1.25	4.65	17.5	37.5	28.5	1900

(continued overleaf)

Table 3.1 (continued)

RG number	Nominal impedance Z0 (ohms)	Overall diameter (inches)	Velocity factor	Attenuation (dB per 100 ft)					Capacity (pF/ft)	Maximum operating voltage RMS
				1 MHz	10 MHz	100 MHz	1000 MHz	3000 MHz		
RG-58C/U	50.0	0.195	0.659	0.42	1.4	4.9	24.0	45.0	30.0	1900
RG-59A/U	75.0	0.242	0.659	0.34	1.10	3.40	12.0	26.0	20.5	2300
RG-59B/U	75	0.242	0.66	–	1.1	3.4	12.0	–	21	2300
RG-62A/U	93.0	0.242	0.84	0.25	0.85	2.70	8.6	18.5	13.5	750
RG-83/U	35.0	0.405	0.66	0.23	0.80	2.8	9.6	24.0	44.0	2000
RG-174A/U	50	0.1	0.66	1.83	3.35	11.0	32.0	64.0	30.0	1400
RG-188A/U	50	0.11	0.66	–	–	10.0	30.0	–	–	1200
*RG-213/U	50	0.405	0.66	0.16	0.6	1.9	8.0	–	29.5	5000
†RG-218/U	50	0.870	0.66	0.066	0.2	1.0	4.4	–	29.5	11 000
‡RG-220/U	50	1.120	0.66	0.04	0.2	0.7	3.6	–	29.5	14 000
URM43	50	0.116	0.66	0.396	1.25	3.96	–	–	30.0	2600
URM70	75	0.128	0.66	0.457	1.46	4.63	–	–	20.5	1800
URM76	50	0.116	0.66	0.457	1.46	4.72	–	–	30.0	2600

*Formerly RG8A/U.
†Formerly RG17A/U.
‡Formerly RG19A/U.

Table 3.2 British UR series

UR number	Nominal impedance Z_0 (ohms)	Overall diameter (inches)	Inner conductor (inches)	Capacity (pF/ft)	Maximum operating voltage RMS	Attenuation (dB per 100 ft)				Nearest RG equivalent
						10 MHz	100 MHz	300 MHz	1000 MHz	
43	52	0.195	0.032	29	2750	1.3	4.3	8.7	18.1	58/U
57	75	0.405	0.044	20.6	5000	0.6	1.9	3.5	7.1	11A/U
63*	75	0.855	0.175	14	4400	0.15	0.5	0.9	1.7	
67	50	0.405	7/0.029	30	4800	0.6	2.0	3.7	7.5	213/U
74	51	0.870	0.188	30.7	15 000	0.3	1.0	1.9	4.2	218/U
76	51	0.195	19/0.0066	29	1800	1.6	5.3	9.6	22.0	58C/U
77	75	0.870	0.104	20.5	125 000	0.3	1.0	1.9	4.2	164/U
79*	50	0.855	0.265	21	6000	0.16	0.5	0.9	1.8	
83*	50	0.555	0.168	21	2600	0.25	0.8	1.5	2.8	
85*	75	0.555	0.109	14	2600	0.2	0.7	1.3	2.5	
90	75	0.242	0.022	20	2500	1.1	3.5	6.3	12.3	59B/U

All the above cables have solid dielectric with a velocity factor of 0.66 with the exception of those marked with an asterisk, which are helical membrane and have a velocity factor of 0.96.

Table 3.3 Foam dielectric, solid outer conductor co-axial RF cables 50 ohms characteristic impedance

Cable type	Superflexible FS-J series		LDF series	
	FSJ1-50A 1/4″	FSJ4-50B 1/2″	LDF2-50 3/8″	LDF4-50A 1/2″
Min. bending radius, in. (mm)	1.0 (25)	1.25 (32)	3.75 (95)	5.0 (125)
Propagation velocity, %	84	81	88	88
Max. operating frequency, GHz	20.4	10.2	13.5	8.8
Att. dB/100 ft (dB/100 m)				
50 MHz	1.27 (4.17)	0.732 (2.40)	0.75 (2.46)	0.479 (1.57)
100 MHz	1.81 (5.94)	1.05 (3.44)	1.05 (3.44)	0.684 (2.24)
400 MHz	3.70 (12.1)	2.18 (7.14)	2.16 (7.09)	1.42 (4.66)
1000 MHz	6.00 (19.7)	3.58 (11.7)	3.50 (11.5)	2.34 (7.68)
5000 MHz	14.6 (47.9)	9.13 (30.0)	8.80 (28.9)	5.93 (19.5)
10 000 MHz	21.8 (71.5)	15.0 (49.3)	13.5 (44.3)	N/A N/A
Power rating, kW at 25°C (77°F)				
50 MHz	1.33	3.92	2.16	3.63
100 MHz	0.93	2.74	1.48	2.53
400 MHz	0.452	1.32	0.731	1.22
1000 MHz	0.276	0.796	0.445	0.744
5000 MHz	0.11	0.30	0.17	0.29
10 000 MHz	0.072	0.19	0.12	N/A

The claimed advantages of foam dielectric, solid outer conductor cables are:

1. Lower attenuation
2. Improved RF shielding, approximately 30 dB improvement
3. High average power ratings because of the improved thermal conductivity of the outer conductor and the lower attenuation.

A disadvantage is that they are not so easy to handle as braided cables.

3.5.3 Voltage standing wave ratio (VSWR)

When an RF cable is mismatched, i.e. connected to a load of a different impedance to that of the cable, not all the power supplied to the cable is absorbed by the load. That which does not enter the load is reflected back down the cable. This reflected power adds to the incident voltage when they are in phase with each other and subtracts from the incident voltage when the two are out of phase. The result is a series of voltage – and current – maxima and minima at half-wavelength intervals along the length of the line (*Figure 3.4*). The maxima are referred to as antinodes and the minima as nodes.

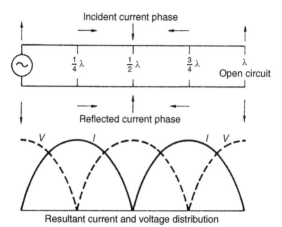

Figure 3.4 Formation of standing waves

The voltage standing wave ratio is the numerical ratio of the maximum voltage on the line to the minimum voltage: VSWR = V_{max}/V_{min}. It is also given by: VSWR = R_L/Z_0 or Z_0/R_L (depending on which is the larger so that the ratio is always greater than unity) where R_L = the load resistance.

The return loss is the power ratio, in dB, between the incident (forward) power and the reflected (reverse) power.

The reflection coefficient is the numerical ratio of the reflected voltage to the incident voltage.

The VSWR is 1, and there is no reflected power, whenever the load is purely resistive and its value equals the characteristic impedance of the line. When the load resistance does not equal the line impedance, or the load is reactive, the VSWR rises above unity.

A low VSWR is vital to avoid loss of radiated power, heating of the line due to high power loss, breakdown of the line caused by high voltage stress, and excessive radiation from the line. In practice, a VSWR of 1.5:1 is considered acceptable for an antenna system, higher ratios indicating a possible defect.

3.5.4 Transmission line filters, baluns and matching circuits

Use can be made of standing waves on sections of line to provide filters and RF transformers. When a line one-quarter wavelength long (aλ/4 stub) is open circuit at the load end, i.e. high impedance, an effective short-circuit is presented to the source at the resonant frequency of the section of line, producing an effective band stop filter. The same effect would be produced by a short-circuited λ/2 section. Unbalanced co-axial cables with an impedance of 50 Ω are commonly used to connect VHF and UHF base stations to their antennas although the antennas are often of a different impedance and balanced about ground. To match the antenna to the feeder and to provide a balance to unbalance transformation (known as a balun), sections of co-axial cable are built into the antenna support boom to act as both a balun and an RF transformer.

Balun

The sleeve balun consists of an outer conducting sleeve, one quarter-wavelength long at the operating frequency of the antenna, and connected to the outer conductor of the co-axial cable as in *Figure 3.5*. When viewed from point Y, the outer conductor of the feeder cable and the sleeve form a short-circuited quarter-wavelength stub at the operating frequency and the impedance between the two is very high. This effectively removes the connection to ground for RF, but not for DC, of the outer conductor of the feeder cable permitting the connection of the balanced antenna to the unbalanced cable without short-circuiting one element of the antenna to ground.

Figure 3.5 Sleeve balun

RF transformer

If a transmission line is mismatched to the load variations of voltage and current, and therefore impedance, occur along its length (standing waves). If the line is of the correct length an inversion of the load impedance appears at the input end. When a λ/4 line is terminated in other than its characteristic impedance an impedance transformation takes place. The impedance at the source is given by:

$$Z_s = \frac{Z_0{}^2}{Z_L}$$

where
Z_s = impedance at input to line
Z_0 = characteristic impedance of line
Z_L = impedance of load

By inserting a quarter-wavelength section of cable having the correct characteristic impedance in a transmission line an antenna of any impedance can be matched to a standard feeder cable for a particular design frequency. A common example is the matching of a folded dipole of 300 Ω impedance to a 50 Ω feeder cable.

Let $Z_s = Z_0$ of feeder cable and $Z_0' =$ characteristic impedance of transformer section. Then:

$$Z_0 = \frac{Z_0'{}^2}{Z_L}$$

$$Z_0' = \sqrt{Z_0 Z_L}$$

$$= \sqrt{300 \times 50} = 122\,\Omega$$

3.6 Waveguides

At the higher microwave frequencies waveguides which conduct electromagnetic waves, not electric currents, are often used. Waveguides are conductive tubes, either of rectangular, circular or elliptical section which guide the wave along their length by reflections from the tube walls. The walls are not used as conducting elements but merely for containment of the wave. Waveguides are not normally used below about 3 GHz because their cross-sectional dimensions must be comparable to a wavelength at the operating frequency. The advantages of a waveguide over a co-axial cable are lower power loss, low VSWR and a higher operating frequency, but they are more expensive and difficult to install.

In a rectangular waveguide an electromagnetic wave is radiated from the source at an angle to the direction of propagation and is

H plane

E plane

Propagation (pre-reflection)

Figure 3.6 Propagation in rectangular waveguide

bounced off the walls (*Figure 3.6*). If the wave were transmitted directly along the length of the guide the electric field would be parallel to one of the walls and be short-circuited by it. Radiating the wave at an angle to the walls creates the maximum field at the centre of the guide and zero at the walls, if the dimensions of the guide are correct for the frequency. However, because the wave does not travel directly along the length of the guide, the speed of propagation is less than in space.

In an electro-magnetic wave the electric and magnetic fields, and the direction of propagation, are mutually perpendicular (see *Figure 1.3*) and such a wave may therefore be thought of as transverse electro-magnetic (TEM). In a waveguide though, because of the short-circuiting effect of the walls, a TEM wave cannot exist. A method of making the wave either transverse electric or transverse magnetic is needed.

When a wave is propagated by a reflection either the magnetic or the electric field is changed. The changed field will now contain the normal component perpendicular to the direction of propagation and a component in its direction, i.e. the wave is no longer wholly transverse. It must be either transverse electric or transverse magnetic. The terminology used to distinguish the type of wave differs: the American system uses the field which behaves as it would in free space to describe the type of wave, e.g. when there is no electric field in the direction of propagation the wave is called TE and the mode with no magnetic field in the direction of propagation, TM; the European system uses the field that is modified and an American TE wave is a European TM wave. In the European system H and E may also be used in lieu of (American) TE and TM.

The behaviour of waves in circular waveguides is similar to that in rectangular guides. Circular waveguides minimize feeder attenuation and are particularly suitable for long vertical runs. A single circular waveguide can carry two polarizations with a minimum isolation of 30 dB. Circular waveguides are recommended where attenuation is critical or where multi-band capability is needed.

Elliptical waveguides have the advantages of flexibility, the availability of long continuous runs and reduced cost.

3.7 Other transmission line considerations

3.7.1 Noise factor of coaxial cable transmission line

Any lossy electrical device, including coaxial cable, produces a noise level of its own. The noise factor of coaxial line is:

$$F_{N(COAX)} = 1 + \frac{(L - 1)T}{290}$$

where
$F_{N(COAX)}$ is the noise factor of the coax
L is the loss expressed as a linear quantity
T is the physical temperature of the cable in Kelvins

The linear noise factor due to loss can be converted to the noise figure by $10\log(F_{N(COAX)})$, which can be added to the system noise decibel for decibel.

The attenuation loss figure published in manufacturers' tables is called the *matched line loss* (L_M) because it refers to the situation where the load and characteristic impedance of the line are equal. But we also have to consider the *Total Line Loss* (TLL), which is:

$$TLL = 10\log\left[\frac{B^2 - C^2}{B(1 - C^2)}\right]$$

where
$B = $ antilog L_M
$C = (SWR_{LOAD} - 1)/(SWR_{LOAD} + 1)$
SWR_{LOAD} is the VSWR at the load end of the line

3.7.2 Types of coaxial cable

Coaxial cable consists of two cylindrical conductors sharing the same axis (hence 'co-axial') and separated by a dielectric. For low frequencies (in flexible cables) the dielectric may be polyethylene or polyethylene foam, but at higher frequencies Teflon® and other materials are used. Also used in some applications, notably high powered broadcasting transmitters, are dry air and dry nitrogen.

Several forms of coaxial line are available. Flexible coaxial cable discussed earlier in this chapter is perhaps the most common form. The outer conductor in such cable is made of either braided wire or foil. Again, television broadcast receiver antennas provide an example of such cable from common experience. Another form of flexible or semi-flexible coaxial line is *helical line* (*Figure 3.7*) in which the outer conductor is spiral wound. This type of coaxial cable is usually 2.5 or more centimetres in diameter.

Figure 3.7 The helical line

Hardline is coaxial cable that uses a thin-walled pipe as the outer conductor. Some hardline coax used at microwave frequencies has a rigid outer conductor and a solid dielectric. *Gas-filled line* is a special case of hardline that is hollow (*Figure 3.8*), the centre conductor

Figure 3.8 Gas-filled line

being supported by a series of thin ceramic or Teflon insulators. The dielectric is either anhydrous (i.e. dry) nitrogen or some other inert gas.

Some flexible microwave coaxial cable uses a solid 'air-articulated' dielectric (*Figure 3.9*), in which the inner insulator is not continuous around the centre conductor, but rather is ridged. Reduced dielectric losses increase the usefulness of the cable at higher frequencies. Double shielded coaxial cable (*Figure 3.10*) provides an extra measure of protection against radiation from the line, and EMI from outside sources from getting into the system.

Figure 3.9   Solid 'air-articulated' dielectric

Figure 3.10   Double shielded coaxial cable

3.7.3 Transmission line noise

Transmission lines are capable of generating noise and spurious voltages that are seen by the system as valid signals. Several such sources exist. One source is coupling between noise currents flowing in the outer and inner conductors. Such currents are induced by nearby electromagnetic interference and other sources (e.g. connection to a noisy ground plane). Although coaxial design reduces noise pick-up compared with parallel line, the potential for EMI exists. Selection of high-grade line, with a high degree of shielding, reduces the problem.

Another source of noise is thermal noise in the resistances and conductances of the line. This type of noise is proportional to resistance and temperature.

There is also noise created by mechanical movement of the cable. One species results from movement of the dielectric against the two conductors. This form of noise is caused by electrostatic discharges in much the same manner as the spark created by rubbing a piece of plastic against woollen cloth.

A second species of mechanically generated noise is piezoelectricity in the dielectric. Although more common in cheap cables, one should be aware of it. Mechanical deformation of the dielectric causes electrical potentials to be generated.

Both species of mechanically generated noise can be reduced or eliminated by proper mounting of the cable. Although rarely a problem at lower frequencies, such noise can be significant at microwave frequencies when signals are low.

3.7.4 Coaxial cable capacitance

A coaxial transmission line possesses a certain capacitance per unit of length. This capacitance is defined by:

$$C = \frac{24\varepsilon}{\log(D/d)} \frac{\text{pF}}{\text{Metre}}$$

where
C is the capacitance
D is the outside conductor diameter
d is the inside conductor diameter
ε is the dielectric constant of the insulator.

A long run of coaxial cable can build up a large capacitance. For example, a common type of coax is rated at 65 pF/metre. A 150 metre roll thus has a capacitance of (65 pF/m) (150 m), or 9750 pF. When charged with a high voltage, as is done in performing breakdown voltage tests at the factory, the cable acts like a charged high voltage capacitor. Although rarely if ever lethal to humans, the stored voltage in new cable can deliver a nasty electrical shock and can irreparably damage electronic components.

3.7.5 Coaxial cable cut-off frequency (F_c)

The normal mode in which a coaxial cable propagates a signal is as a transverse electromagnetic (TEM) wave, but others are possible – and usually undesirable. There is a maximum frequency above which TEM propagation becomes a problem, and higher modes dominate. Coaxial cable should not be used above a frequency of:

$$F_{\text{CUT-OFF}} = \frac{1}{3.76(D + d)\sqrt{\varepsilon}}$$

where
F is the TEM mode cut-off frequency
D is the diameter of the outer conductor in mm

d is the diameter of the inner conductor in mm

ε is the dielectric constant

When maximum operating frequencies for cable are listed it is the TEM mode that is cited. Beware of attenuation, however, when making selections for microwave frequencies. A particular cable may have a sufficiently high TEM mode frequency, but still exhibit a high attenuation per unit length at X or Ku-bands.

References

Andrew Antennas, (1991). *Catalogue 35*, Illinois.

British Telecommunications (1992). *Connect Direct, private circuits*. BT, London.

Kennedy, G. (1977). *Electronic Communications Systems*. McGraw-Hill Kogashuka, Tokyo.

Terman, F.E. (1943), *Radio Engineers' Handbook*, McGraw-Hill, London.

Winder, S.W. (2001). *Newnes Telecommunications Pocket Book*. Butterworth-Heinemann, Oxford.

4 Antennas

4.1 Antenna characteristics

4.1.1 Bandwidth

Stated as a percentage of the nominal design frequency, the bandwidth of an antenna is the band of frequencies over which it is considered to perform acceptably. The limits of the bandwidth are characterized by unacceptable variations in the impedance which changes from resistive at resonance to reactive, the radiation pattern, and an increasing VSWR.

4.1.2 Beamwidth

In directional antennas the beamwidth, sometimes called half-power beamwidth (HPBW), is normally specified as the total width, in degrees, of the main radiation lobe at the angle where the radiated power has fallen by 3 dB below that on the centre line of the lobe (*Figure 4.1A*).

Figure 4.1A   Half-power beamwidth

4.1.3 Directivity and forward gain

All practical antennas concentrate the radiated energy in some directions at the expense of others. They possess directivity but are completely passive; they cannot increase the power applied to them. Nevertheless, it is convenient to express the enhanced radiation in some directions as a power gain.

Antenna gain may be quoted with reference to either an isotropic radiator or the simplest of practical antennas, the dipole. There is a difference of 2.15 dB between the two figures. A gain quoted in dBi is with reference to an isotropic radiator and a gain quoted in dBd is with reference to a dipole. When gain is quoted in dBi, 2.15 dB must be subtracted to relate the gain to that of a dipole.

4.1.4 Effective height or length

The current flowing in an antenna varies along its length (see *Figure 1.4*). If the current were uniform along the length of an antenna it would produce a field appropriate to its physical length, and the effective height or length of the antenna would be its physical length. In practice, because the current is not uniform, the effective length is less than the physical length and is given by:

$$l_{\text{eff}} = \frac{l_{\text{phys}} \times I_{\text{mean}}}{I}$$

where
l_{eff} = effective length
l_{phys} = physical length
I = current at feed point

With an antenna which is short in comparison with a wavelength the current can be considered to vary linearly over its length and $I_{\text{mean}} = I/2$. Because the apparent length of a vertical radiator is twice that of its physical dimension due to the mirror image formed below the ground, the effective length of an electrically short vertical antenna may be approximated to be its physical length.

4.1.5 Effective radiated power (erp)

This is the power effectively radiated along the centre line of the main lobe. It is the power supplied to the antenna multiplied by the antenna gain with reference to a dipole.

4.1.6 Radiation resistance and efficiency

The power radiated by an antenna can conveniently be expressed in terms of the value of a resistor which would dissipate the same power that the antenna radiates. This value is referred to as the radiation resistance and is defined as the ratio of the power radiated to the square of the current at the feed point. The efficiency is the ratio of the power radiated to that lost in the antenna. It is given by:

$$\text{eff} = \frac{R_r}{R_r + R_L} \times 100\%$$

where R_r is the radiation resistance and R_L represents the total loss resistance of the antenna. The sum of the two resistances is the total resistance of the antenna and, for a resonant antenna, is also the impedance.

4.1.7 Front-to-back ratio

The ratio, in dB, of the strength of the radiation (or received signal) in the forward (desired) direction to that in the reverse (unwanted) direction. The front-to-back ratio of the antenna shown in *Figure 4.1A* is 13 dB.

4.1.8 Impedance

The impedance of an antenna is that presented to the feeder cable connecting it to the transmitter or receiver. It is the result of the vectorial addition of the inductive, capacitive and resistive elements of the antenna. Each resonant antenna possesses an impedance characteristic of the type, and when an antenna operates at its resonant frequency the reactive elements cancel out and the impedance becomes resistive. The radiation resistance plus the losses in the antenna, i.e. the series resistance of the conductors, the shunt resistance of the base material and losses in nearby objects, form the resistive portion of the impedance.

4.1.9 Polarization

The radiated field from an antenna is considered to be polarized in the plane of the length of the conductors which is the plane of the electric field, the E plane. Confusion arises when reference is made to vertical or horizontal polarization and it is preferable when referring to polar diagrams to use the E and H plane references.

Circular polarization, produced by crossed dipoles or helical wound antennas, is occasionally used for point-to-point work at VHF and above to reduce multi-path propagation losses.

Cross polarization discrimination defines how effectively an antenna discriminates between a signal with the correct polarization, i.e. mounted with the elements in the same plane, and one operating at the same frequency with the opposite polarization. 20 dB is typical.

4.1.10 Radiation pattern

A plot of the directivity of an antenna showing a comparison of the power radiated over 360°. Two polar diagrams are required to show the radiation in the E and H planes. The polar diagrams may be calibrated in either linear (voltage) or logarithmic (decibel) forms.

4.1.11 Voltage standing wave ratio (VSWR)

Most VHF and UHF antennas contain an impedance matching device made up of lengths of co-axial cable. Thus the VSWR (see Chapter 3) of these types of antenna varies with the operating frequency, more so than the bandwidth of the antenna alone would produce. At the centre design frequency, the VSWR should, theoretically, be 1:1 but in practice a VSWR less than 1.5:1 is considered acceptable.

4.1.12 Receive aperture

Receiving antennas also possess a property called aperture, or capture area. This concept relates the amount of power that is delivered to a matched receiver to the power density (watts per square metre). The aperture is often larger than the physical area of the antenna, as in the case of the half-wavelength dipole (where the wire fronts a very small physical area), or less as in the case of a parabolic reflector used in microwave reception. *Figure 4.1B* shows the capture area of a half-wavelength (0.5λ) dipole. It consists of an ellipse with major axes of 0.51λ and 0.34λ. The relationship between gain and aperture is:

$$A_e = \frac{G\lambda^2}{4\pi n}$$

where
A_e is the effective aperture
G is the gain
λ is the wavelength of the signal
n is the aperture effectiveness ($n = 1$ for a perfect no-loss antenna, real values are typically 0.3 to 0.55

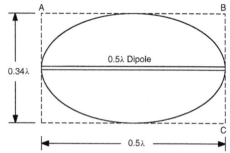

Figure 4.1B   Capture area of half-wavelength (0.5λ) dipole

4.2 Antenna types

4.2.1 The dipole

The half-wavelength (λ/2) dipole as described in Section 1.4 is the antenna on which many others are based. *Figure 4.2* shows the relative radiation in the E and H planes of a dipole in free space.

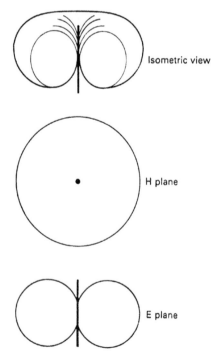

Figure 4.2   Radiation patterns of a half-wave dipole

The impedance of a half-wavelength dipole is 72 Ω; that of a full wavelength or folded dipole is 300 Ω.

4.2.2 The quarter-wavelength vertical radiator

The quarter-wavelength (λ/4) vertical radiator is a commonly used antenna for MF broadcasting and for VHF and UHF mobile radio applications. It is derived from the λ/2 dipole and it is assumed that a mirror image of the radiator is formed below the ground to complete the λ/2 structure of the dipole as in *Figure 4.3*. The radiation pattern of a λ/4 vertical radiator mounted close to a perfect earth shows a strong similarity to that of a dipole. The effect of reducing the size and conductivity of the ground plane raises the angle of radiation.

Figure 4.3   Quarter-wavelength vertical radiator

The impedance of a perfect λ/4 vertical radiator is 36 Ω but reducing the effectiveness of the ground plane raises the impedance.

4.2.3 LF, MF and HF antennas

Because of the physical lengths involved, LF and MF antennas are usually non-resonant and their impedances do not conform to the resistive 70 Ω or 36 Ω of the basic resonant types. The impedance of a non-resonant antenna is usually higher and reactive so an antenna tuning or matching unit is used to couple the antenna efficiently to the transmission line and also act as filter to reduce out-of-band radiations. The matching unit comprises a tuned circuit with either a tap on the coil at the correct impedance point or a separate coupling coil to feed the antenna.

Obtaining an adequate length is always the problem with low frequency antennas and various methods have been used based mainly on the λ/4 radiator. The horizontal section of the inverted L (*Figure 4.4*) extends the effective length but, as the ground wave is much used at

Figure 4.4 Inverted L antenna

the lower frequencies, these antennas are intended for vertical polarization and it is therefore only the down-lead which radiates, or receives, effectively. An alternative method of increasing the effective height of a vertical radiator is to provide a capacitance top where the system of horizontal conductors provides a high capacitance to ground. This prevents the current falling to zero at the top of the antenna, maintaining a higher mean current and so increasing the antenna's effective length.

Dipoles used at HF are mounted horizontally because of their length and have directivity in the horizontal (E) plane. Propagation is mainly by the sky wave and the omni-directional properties in the vertical (H) plane, modified by ground reflections, produce wide angle upwards radiation.

4.2.4 Directional arrays

Broadside array

A broadside array consists of several radiators spaced uniformly along a line, each carrying currents of the same phase. When each radiator has an omni-directional pattern, and the spacing between radiators is less than $3\lambda/4$, maximum radiation occurs at right angles to the line of the array. The power gain is proportional to the length of the array, provided that the length is greater than two wavelengths; this means, effectively, the number of radiators. *Figure 4.5* shows a typical H plane polar diagram for an array with vertically mounted radiators and a spacing of $\lambda/2$.

End-fire array

Physically an end-fire array is identical to a broadside except for the feeding arrangements and the spacing of the elements. In an end-fire array the radiators are fed with a phase difference between adjacent radiators equal in radians to the spacing between them in wavelengths.

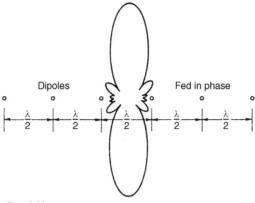

Figure 4.5 Broadside array

A spacing of λ/4 requires a phase shift of 90° between adjacent radiators. *Figure 4.6* shows a typical radiation pattern. An end-fire array concentrates the power in both the E and H planes and the maximum radiation is in the direction of the end of the array with the lagging phase.

Figure 4.6 End-fire array

Rhombic antenna

A rhombic is a wide band, directional antenna comprised of four non-resonant wire antennas, each several wavelengths long, arranged as shown in *Figure 4.7(a)* which also shows the radiation pattern for each leg of the rhombus. The lobe angle θ can be varied by adjusting the length, in wavelengths, of each radiator. The antenna has greater directivity than a single wire and can be terminated by an appropriate value resistor to ensure non-resonance and a wide bandwidth. However, because it must be terminated in a resistance equal to the characteristic impedance of the conductors, it cannot be more than 50% efficient. It also exhibits considerable side lobes of radiated power. Rhombics are used for sky wave working at HF and more than one frequency is allocated to allow for varying propagation conditions. The conductors of a rhombic are normally horizontal and the horizontal directivity is

60

determined by the tilt angle, β in *Figure 4.7(a)*. If the lobe angle θ is equal to $(90 - \beta)°$ the radiation in the A lobes cancels, while that from the B lobes, which point in the same direction, is added. The resultant pattern in the horizontal plane is shown in *Figure 4.7(b)*. The vertical directivity is controlled by the height of the conductors above the ground.

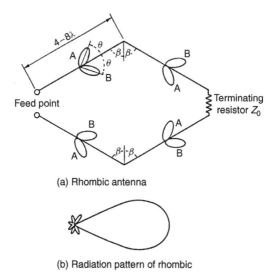

(a) Rhombic antenna

(b) Radiation pattern of rhombic

Figure 4.7

Log-periodic antenna

An alternative, usable from HF through UHF, to the rhombic for wide band operation is the log-periodic antenna. It is comprised of several dipoles of progressive lengths and spacings as in *Figure 4.8*, and is resonant over a wide frequency range and may be mounted with either polarization. The dipoles are fed via the support booms and this construction ensures that the resultant phasing of the dipoles is additive in the forward direction producing an end-fire effect. However, because at any one frequency only a few of the dipoles are close to resonance, the forward gain of the antenna is low considering the number of elements it contains.

Figure 4.8 Log-periodic antenna at VHF frequency

4.3 VHF and UHF antennas

4.3.1 Base station antennas

Apart from entertainment broadcasting and mobile telephony, most VHF and UHF systems use vertical polarization and a dipole – or to prevent noise from rain static, the folded dipole – with the conductors mounted vertically is a frequently used antenna for VHF and UHF base station installations. Unfortunately it is often mounted on the side of the support structure in a manner which seriously affects its omnidirectional radiation pattern. Where practical, there should be a minimum spacing of one wavelength between the structure and the rearmost element of the antenna.

To obtain a good omni-directional pattern either an end-fed dipole (*Figure 4.9*) or a unipole antenna (*Figure 4.10*) protruding from the top of the mast or tower is the best option. A unipole is a variation of the vertical quarter-wave radiator and provides a low angle of radiation.

To reduce the likelihood of co-channel interference directional antennas are often necessary. The simplest of these is the combination of a λ/2 dipole and reflector shown in *Figure 4.11*. The reflector is slightly longer than the dipole and spaced one quarter-wavelength from it. The portion of the signal radiated by the dipole in the direction of the reflector is received and re-transmitted by the reflector, with a 180° phase change occurring in the process. The signal re-transmitted

Figure 4.9 Type EDV end-fed antenna (by kind permission of C and S Antennas Ltd)

to the rear of the antenna – the direction of the reflector – cancels the signal from the dipole, that towards the front of the antenna adds to the signal from the dipole giving the radiation pattern shown. The power gain of a dipole and reflector, a two-element array, is 3 dBd.

Directivity can be increased by adding directors forward of the dipole, the result is a Yagi–Uda array. The limit to the number of radiators is set by physical constraints and the reduction of bandwidth produced by their addition. At low VHF, a 3-element array is about the practical maximum, while at 1500 MHz, 12-element arrays are commonplace. As a rule of thumb, doubling the number of elements in an array increases the forward gain by 3 dB. Where the maximum front-to-back ratio is essential the single rod reflector can be replaced by a corner reflector screen.

Co-linear antennas provide omni-directional characteristics and power gain in the H plane. A co-linear consists of a number of dipoles stacked vertically and, in the normal configuration, fed so that they radiate in phase and the maximum power is radiated horizontally. *Figure 4.12* shows alternative feeding arrangements. One advantage of the co-linear is that the horizontal angle of radiation can be tilted to about 15° downwards by changing the phasing of the elements. The gain of a co-linear is limited, because of the physical lengths involved and losses in the feeding arrangements to 3 dBd at VHF and 6 dBd at UHF.

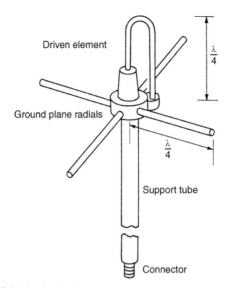

Figure 4.10 Folded unipole antenna

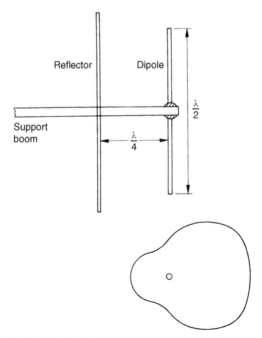

Figure 4.11 Dipole and reflector

64

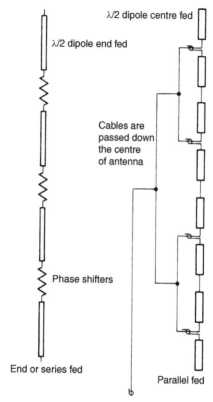

λ/2 dipole centre fed

λ/2 dipole end fed

Cables are
passed down
the centre
of antenna

Phase shifters

End or series fed

Parallel fed

Figure 4.12 General construction of co-linears

Figure 4.13 shows a slot antenna cut into a flat metal sheet. Current (I) injected at the centre of the slot flows around the edge and creates a vertical electric field. The radiated field pattern is like a dipole.

The type of slot antenna typically used for mobile telephony base stations is a cylindrical waveguide with slots cut width-wise. Current flowing along the waveguide creates an electric field along the length of the cylinder. The radiation pattern produced by a slot antenna cut into a cylinder is directional, with the main beam perpendicular to the slot. Using two slot antennas side by side provides radio coverage over a 120° sector. Three pairs of slot antennas placed around a mast gives three sectors that can operate at different frequencies.

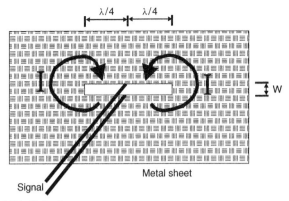

Figure 4.13 Slot antenna

A wide-band alternative to the log-periodic is the conical (discone) antenna (*Figure 4.14*). It provides unity gain, is omni-directional and has a bandwidth of approximately 3:1, depending on the designed frequency range. In practice there has been a tendency to expect these antennas to perform outside their specified bandwidths with unsatisfactory results.

Stacking and baying

A method of increasing an antenna's directivity is to mount two or more antennas vertically above one another (stacking) or side-by-side (baying), and to feed them so that they radiate in phase. Stacking two dipoles vertically increases the directivity in the E plane and baying them increases the directivity in the H plane, approximately halving the beamwidth in each case.

An array of two stacked plus two bayed antennas approximately halves the beamwidth in both planes.

4.3.2 Mobile antennas

The aerial is the least expensive, and most abused, component of a mobile radio installation. Frequently installed in a manner which does not produce optimum performance it can have a profound effect on the performance of the whole installation.

Most mobile antennas consist of a metal rod forming a quarter-wavelength radiator. The ideal mounting position is the centre of a

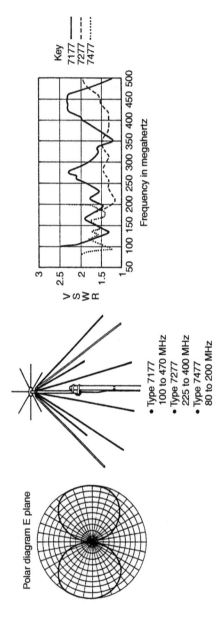

Polar diagram E plane

- Type 7177
  100 to 470 MHz
- Type 7277
  225 to 400 MHz
- Type 7477
  80 to 200 MHz

Key

7177 ———
7277 – – –
7477 ·······

Figure 4.14 Discone wide band antenna (by kind permission of Jaybeam Ltd) metallic roof, and as the area of the ground plane is reduced the radiation pattern changes and more of the energy is radiated upwards (not always a bad thing in inner city areas); also, the impedance rises.

The effect of the mounting position on the H plane radiation can be dramatic, resulting in ragged radiation patterns and, in some directions, negligible radiation. Advice on the installation of mobile antennas and the polar diagrams produced by typical installations are illustrated in MPT 1362, *Code of Practice for Installation of Mobile Radio Equipment in Land Based Vehicles.*

As the installation moves away from the ideal and the antenna impedance rises a mismatch is introduced between the antenna and the feeder with the consequent production of standing waves on the feeder. Under high VSWR conditions the cable is subject to higher voltage stresses and it also behaves as an aerial radiating some of the reflected power. This spurious radiation adds to the radiation from the antenna in some directions but subtracts from it in others giving rise to jagged radiation patterns or deep nulls in radiated signal.

Mobile antennas providing a small amount of gain, typically 3 dB and obtained by narrowing the radiation lobes, are on the market. These have a length of 5/8 wavelength and, because the extra length makes the impedance capacitive at the operational frequency, a loading coil is inserted at the lower end of the element to cancel the capacitive reactance. An adjustable metallic disk towards the base of the whip is often provided for tuning purposes. Note that gain figures quoted for mobile antennas are usually with reference to a quarter-wave whip.

Low profile antennas

Low profile antennas are available for use at UHF. They have a built-in ground plane approximately 150 mm in diameter and a height of some 30 mm and have obvious applications for use on high vehicles and, although not strictly covert, where a less obtrusive antenna is required. They are fitted with a tuning screw and when adjusted to resonance a VSWR of better than 1.2:1 is quoted by one maker and a bandwidth of 10 MHz at a VSWR of 2:1. *Figure 4.15* shows the radiation pattern for one type.

Motor-cycle antennas

The installations of antennas on motor cycles poses problems because of the absence of a satisfactory ground plane. One frequently used method is to employ a 5/8 wavelength whip and loading coil. Another

68

Tunable range .... 400–475 MHz

Bandwidth at: 3:1 VSWR ..... 15 MHz
2:1 VSWR ..... 10 MHz

VSWR at resonance .... < 1.2:1

Typical radiation pattern
(460 MHz)

Antenna mounted on a 40 cm × 40 cm aluminium plate

Figure 4.15 Low profile UHF antenna (by kind permission of Panorama Antennas Ltd)

method uses a pair of grounded downwards-pointing rods to form the lower half of a dipole.

Hand-portable antennas

Again, because of the lack of a ground plane high performance antennas are difficult to provide for hand-portables, particularly at VHF. Body-worn sets may have an antenna incorporated in the microphone lead but the high current portion of the antenna must then be at a low height and in some directions the radiation must pass through the body, which is highly absorbent, to reach the base station. Helical antennas are frequently used on hand-held sets to reduce the physical length. Useful operating tips are to face the base station when using the radio in low signal areas, while placing the set on a nearby car roof effectively increases the performance of the antenna.

Safety

There are two safety aspects to consider when installing mobile antennas: physical and electrical. The physical considerations are that the antenna must be incapable of inflicting injury when it is in its correct position, and also when it has been bent or damaged. Rear wing mounted antennas need particular care in their positioning; a Band 111 aerial tip is just about eye height when bending over an open boot lid. The same considerations apply to hand-portables, helical antennas being perhaps safer than whips because they are thicker and thus more easily seen. They also have rounded tips.

The electrical dangers are from radiation affecting the body either directly – radiation from a hand-portable helical into the eye is a possible example – or indirectly by affecting electronic equipment. The danger of radiation affecting equipment in the vehicle is increased when the VSWR is high because of increased radiation from the feeder. Advice should be sought from the Radiological Protection Board.

4.4 Microwave antennas

The small antenna elements at microwaves facilitate the construction of highly directive, high gain antennas with high front-to-back ratios.

At frequencies below about 2 GHz, 12- to 24-element Yagi arrays, enclosed in plastic shrouds for weather protection, may be used. At higher frequencies, antennas with dish reflectors are the norm.

The aperture ratio (diameter/wavelength) of a dish governs both its power gain and beamwidth. The power gain of a parabolic dish is given to a close approximation by:

$$\text{Gain} = 10 \log_{10} 6(D/\lambda)^2 \times N, \text{dBi}$$

where $D =$ dish diameter and $N =$ efficiency. Dimensions are in metres. The half-power beam width (HPBW) in degrees is approximately equal to $70\lambda/D$.

A microwave antenna with its dish reflector, or parasitic elements in the case of a Yagi type, is a large structure. Because of the very narrow beamwidths – typically $5°$ for a 1.8 m dish at 2 GHz – both the antenna mounting and its supporting structure must be rigid and able to withstand high twisting forces to avoid deflection of the beam in high winds. Smooth covers, radomes, fitted to dishes and the fibreglass shrouds which are normally integral with Yagis designed for these applications considerably reduce the wind loading and, for some antenna types, increase the survival wind speed.

The electrical performance of a selection of microwave antennas is given in *Table 4.1* and the wind survival and deflection characteristics in *Table 4.2* (Andrew Antennas, 1991).

Table 4.1 2.1–2.2 GHz antennas – electrical characteristics

Type number	Dia. (m)	Gain (dBi)			Beam width (deg.)	Cross pol. disc (dB)	F/B ratio (dB)	VSWR max.
		Bottom	Mid-band	Top				
Ultra High Performance Antenna, F-Series Unpressurized – Radome Inc.								
Single polarized								
UHP8F-21	2.4	31.9	32.1	32.3	4.2	32	61	1.10
UHP10F-21	3.0	33.7	33.9	34.0	3.6	33	64	1.10
UHP12F-21	3.7	35.4	35.6	35.8	2.9	32	65	1.10
Dual polarized								
UHX8F-21	2.4	31.9	32.1	32.3	4.2	30	58	1.20
UHX10F-21A	3.0	33.8	34.0	34.2	3.6	32	62	1.20
UHX12F-21A	3.7	35.4	35.6	35.8	2.8	32	67	1.20
High Performance Antenna, F-Series Unpressurized – Radome Inc.								
Single polarized								
HP6F-21B	1.8	29.4	29.6	29.8	5.5	30	46	1.12
HP8F-21A	2.4	31.9	32.1	32.3	4.1	30	53	1.12
HP10F-21A	3.0	33.8	34.0	34.2	3.4	32	55	1.12
HP12F-21A	3.7	35.4	35.6	35.8	2.9	32	56	1.12
Standard Antenna, F-Series Unpressurized								
Single polarized								
P4F-21C	1.2	26.4	26.6	26.8	7.6	30	36	1.15
P6F-21C	1.8	29.8	30.0	30.2	4.9	30	39	1.12
P8F-21C	2.4	32.3	32.5	32.7	3.8	30	40	1.12
P10F-21C	3.0	34.0	34.2	34.4	3.4	30	44	1.12
Grid Antenna, F-Series Unpressurized								
Single polarized								
GP6F-21A	1.8	29.6	29.8	30.0	5.4	31	36	1.15
GP8F-21A	2.4	32.0	32.2	32.4	4.0	35	39	1.15
GP10F-21A	3.0	34.0	34.2	34.4	3.3	40	42	1.15
GP12F-21	3.7	35.5	35.7	35.9	2.8	40	44	1.15

With shrouded Yagis and some dishes low loss foam-filled cables are generally used up to about 2 GHz although special connectors may

Table 4.2 Wind survival and deflection characteristics

Antenna types	Survival ratings		Max. deflection in 110 km wind (degrees)
	Wind velocity (km/h)	Radial ice (mm)	
P4F Series			
Without Radome	160	12	0.1
With Radome	185	12	0.1
Standard Antennas (except P4F Series)			
Without Radome	200	25	0.1
With Standard Radome	200	25	0.1
UHX, UMX, UGX Antennas	200	25	0.1

be required. At higher frequencies, air-spaced or pressurized nitrogen-filled cables are frequently used with waveguides as an alternative.

4.4.1 Omnidirectional normal mode helix

The normal mode helix antenna shown in *Figure 4.16* produces an omnidirectional pattern when the antenna is mounted vertically. The diameter (D) of the helical coil should be one-tenth wavelength ($\lambda/10$), while the pitch (i.e. S, the distance between helix loops) is one-twentieth wavelength ($\lambda/20$). An example of the normal mode helix is the 'rubber ducky' antenna used on VHF/UHF two-way radios and scanners.

4.4.2 Axial mode helical antenna

An axial mode helical antenna is shown in *Figure 4.17*. This antenna fires 'off-the-end' in the direction shown by the arrow. The helix is mounted in the centre of a ground plane that is at least 0.8λ across. For some UHF frequencies some manufacturers have used aluminum pie pans for this purpose. The helix itself is made from either heavy copper wire (solid, not stranded) or copper or brass tubing. The copper tubing is a bit easier to work. The dimensions are:

$$D \approx \lambda/3$$

$$S \approx \lambda/4$$

$$\text{Length} \approx 1.44\lambda$$

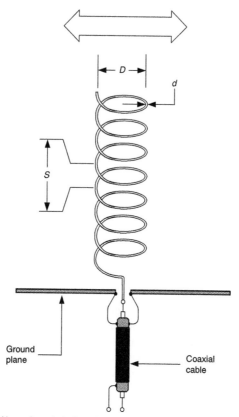

Figure 4.16 Normal mode helix antenna

A 'rule of thumb' for the circumference is that maximum gain is obtained when circumference C is:

$$C = 1.066 + [(N - 5) \times 0.003]$$

4.4.3 Small loop antennas

Small loop antennas are used mostly for receiving, although some designs are also used for transmitting. One application for the small loop antenna is radio direction finding (RDF). Another use is for providing a small footprint antenna for people who cannot erect a full sized receiving antenna. Perhaps the greatest use of the small loop antenna is for receiving stations on crowded radio bands. The small loop antenna has very deep nulls that make it easy to null out interfering co-channel and adjacent channel signals.

Figure 4.17 Axial mode helical antenna

4.5 Loop antennas

4.5.1 Small loop antennas defined

Large loop antennas are those with overall wire lengths of 0.5λ to more than 2λ. Small loop antennas, on the other hand, have an overall wire length that is much less than one wavelength (1λ). According to a Second World War US Navy training manual such antennas are those with an overall length of $\leq 0.22\lambda$. Jasik's classic 1961 text on radio antennas uses the figure $\leq 0.17\lambda$, while John Kraus (1950) used the figure $\leq 0.10\lambda$. An amateur radio source, *The ARRL Antenna Book*, recommends $\leq 0.085\lambda$ for small loop antennas. For the purposes of our discussion we will use Kraus's figure of $\leq 0.10\lambda$.

A defining characteristic of small loops versus large loops is seen in the current distribution. In the small loop antenna the current flowing in the loop is uniform in all portions of the loop. In the large loop, however, the current varies along the length of the conductor, i.e. there are current nodes and antinodes.

The small loop antenna also differs from the large loop in the manner of its response to the radio signal. A radio signal is a transverse electromagnetic (TEM) wave, in which magnetic and electrical fields alternate with each other along the direction of travel. The large loop, like most large wire antennas, respond primarily to the *electrical field* component of the TEM, while small loops respond mostly to the *magnetic field* component. The importance of this fact is that it means the small loop antenna is less sensitive to local electromagnetic interference sources such as power lines and appliances. Local EMI consists largely of electrical fields, while radio signals have both magnetic and electrical fields. With proper shielding, the electrical response can be reduced even further.

4.5.2 Small loop geometry

Small loop antennas can be built in any of several different shapes (*Figure 4.18*). Popular shapes include hexagonal (*Figure 4.18A*), octagonal (*Figure 4.18B*), triangular (*Figure 4.18C*), circular (*Figure 4.18D*) and square (*Figure 4.18E*).

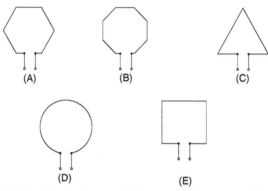

(A)   (B)   (C)

(D)   (E)

Figure 4.18 Small loop antennas: (A) hexagonal, (B) octagonal, (C) triangular, (D) circular and (E) square

The far-field performance of small loop antennas is approximately equal provided that $A^2 \leq \lambda/100$, where A is the loop area and λ is the wavelength of the desired frequency.

The 'standard' loop used in this discussion is the square loop depicted in *Figure 4.19*. There are two forms of winding used: *depth wound* and *spiral wound*. The difference is that the depth wound has its turns in different parallel planes, while in the spiral wound version all of the turns are in the same plane. The spiral wound loop theoretically has a deeper null than depth wound, but in practical terms there is usually little difference.

The length of each side of the loop is designated A, while the depth or width is designated B. These dimensions will be used in equations shortly. The constraint on the dimensions is that A should be $\leq 0.10\lambda/4$ and $B \leq A/5$.

The loop may be either tuned or untuned. The differences between tuned and untuned will be discussed shortly.

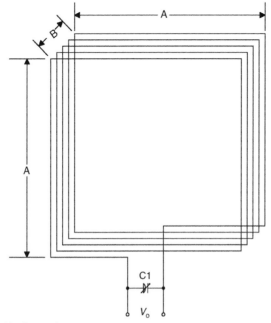

Figure 4.19 Square loop

4.5.3 Small loop antenna patterns

Small loop antennas have patterns opposite those of large loops. The minima, or 'nulls', are perpendicular to the plane of the loop, while the maxima are off the ends. *Figure 4.20* shows the directions of maximum and minimum response. The loop antenna is viewed from above. The nulls are orthogonal to the loop axis, while the maxima are along the loop axis.

The fact that the small loop pattern has nulls perpendicular to the loop axis, i.e. perpendicular to the plane of the loop, is counterintuitive to many people. The advancing radio wave produces alternating regions of high and low amplitude. A *potential difference* exists between any two points. When the loop is aligned such that its axis is parallel to the isopotential lines low signal levels are induced into the loop. If the turns of the loop cut several isopotential lines, a larger signal is induced from this direction.

4.5.4 Signal voltage (V_o) developed by the loop

The actual signal voltage (V_o) at the output of the terminals of an untuned loop is a function of the direction of arrival of the signal (α),

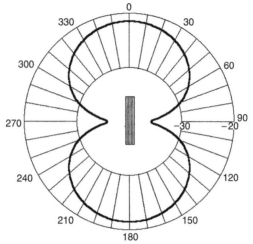

Figure 4.20 Small loop azimuth

as well as the strength of the arriving signal and the design of the loop. The angle α is formed between the loop axis and the advancing isopotential lines of the radio signal.

The output voltage of an untuned loop is:

$$V_\text{o} = \frac{2\pi ANE_\text{f}\cos\alpha}{\lambda}$$

While the output voltage of a tuned loop is:

$$V_\text{o} = \frac{2\pi ANQE_\text{f}\cos\alpha}{\lambda}$$

where
V_o is the output voltage of an untuned loop in volts (V)
A is the length of one side of the loop
N is the number of turns in the loop
E_f is the strength of the incoming signal in volts per metre (V/m)
α is the angle between the advancing wavefront and the loop axis
λ is the wavelength of the radio signal in metres (m), i.e. the reciprocal of the frequency ($\lambda = 1/F$)
Q is the loaded Q (figure of merit) of the tuned circuit formed by C1 and the loop inductance (typically 10 to 100)

4.5.5 Effective height

Loop antennas are sometimes described in terms of the *effective height* (H_{eff}) of the antenna. This number is a theoretical construct that compares the output voltage of a small loop with a vertical piece of the same kind of wire that has a height of:

$$H_{eff} = \frac{2\pi NA}{\lambda}$$

where H_{eff} is the effective height in metres, and all other terms are as defined above.

Loop inductance

A loop antenna essentially forms a coil of wire (or other conductor), so will have inductance. There are several methods for calculating loop inductance, but the most common are the *Grover equation* and the *Patterson equation*.

Grover equation:

$$L_{\mu H} = \left[K_1 N^2 A \text{Ln} \left[\frac{K_2 AN}{(N+1)B} \right] \right] + K_3 + \left[\frac{K_4(N+1)B}{AN} \right]$$

where all terms are as previously defined, except $K_1 - K_4$, which are defined in *Table 4.3*.

Table 4.3 *K* factors for calculating loop inductance

Shape	K_1	K_2	K_3	K_4
Triangle	0.006	1.1547	0.65533	0.1348
Square	0.008	1.4142	0.37942	0.3333
Hexagon	0.012	2.00	0.65533	0.1348
Octagon	0.016	2.613	0.75143	0.07153

Patterson equation:

$$L\mu H = (0.00508A) \times \left[2.303 \log \left(\frac{4A}{d} \right) - \phi \right]$$

where
d is the conductor diameter
ϕ is a factor found in *Table 4.4*

Table 4.4 ϕ Factor for calculating loop inductance

Shape	Factor (ϕ)
Circle	2.451
Octagon	2.561
Hexagon	2.66
Pentagon	2.712
Square	2.853
Triangle	3.197

Of these equations, most people will find that the Grover equation most accurately calculates the actual inductance realized when a practical loop is built.

References

Andrew Antennas (1991). *Catalog 35*. Illinois.

Belcher, R. *et al.* (1989). *Newnes Mobile Radio Servicing Handbook*. Butterworth-Heinemann, Oxford.

Davies, J. (1987). *Private Mobile Radio – A Practical Guide*. Heinemann, London.

Green, D.C. (1979). *Radio systems, Vols II and III*. Pitman, London.

Kennedy, G. (1977). *Electronic Communication Systems*. McGraw-Hill Kogashuka, Tokyo.

Terman, F.E. (1943). *Radio Engineers' Handbook*. McGraw-Hill, London.

5 Resonant circuits

5.1 Series and parallel tuned circuits

Tuned resonant circuits composed of inductance and capacitance are used to generate alternating voltages of a specific frequency and to select a wanted frequency or band of frequencies from the spectrum. *Figure 5.1* contains the diagrams of series and parallel resonant circuits including the resistances which account for the losses present in all circuits. In practice the greatest loss is in the resistance of the inductor, R_L, with a smaller loss, r_c, occurring in the dielectric of the capacitor.

Figure 5.1   Series and parallel resonant circuits

5.1.1 Series resonance

Off resonance, the series circuit exhibits a high impedance to a voltage applied across A and B. This impedance is formed by the vectorial addition of the reactances of the inductance and capacitance at the applied frequency plus the resistances. Ignoring the dielectric losses and the very small stray shunt capacitance C_s, the resonant impedance is given by:

$$Z = \sqrt{R^2 + \left(\omega L - \frac{1}{\omega C}\right)^2}$$

80

where
R = resistance of components in ohms
$\omega = 2\pi \times$ frequency in hertz
L = inductance in henries
C = capacitance in farads

At resonance the reactances cancel out and the impedance falls to approximately the value of the resistance, R, and a maximum line current will flow. *Figure 5.2* shows the response curves of series and parallel circuits near resonance. The resonant frequency is given by:

$$f = \frac{1}{2\pi \sqrt{LC}}$$

where
f = resonant frequency in hertz
L = inductance in henries
C = capacitance in farads

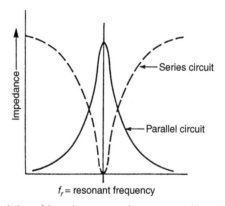

Figure 5.2 Variation of impedance around resonance with series and parallel tuned circuits

5.1.2 Parallel resonance

At resonance, calculated by using the same formula as for a series circuit, the impedance of a parallel tuned circuit is also resistive, and the circulatory current in the circuit is high producing the maximum voltage across the inductance and capacitance. Consequently, at resonance the minimum line current flows. The impedance at resonance or dynamic resistance of a parallel tuned circuit of moderate to high Q is given by:

$$R_d = \frac{L}{CR}$$

where
R_d = dynamic resistance of circuit
R = resistance of components in ohms
L = inductance in henries
C = capacitance in farads

5.2 Q factor

The voltages across the inductor and capacitor in a circuit at resonance are substantially opposite in phase (the loss resistances affect the phasing slightly) and cancel each other. The voltage developed across the inductor, usually the lossiest component, compared with the voltage applied in series to the circuit, is a measure of the 'goodness' of the circuit. This ratio is often referred to as the magnification factor, Q, of the circuit. The Q factor is calculated from:

$$Q = \frac{\omega L}{R} \text{ or } \frac{1}{\omega C R}$$

where L and C are in henries and farads respectively.

5.3 Coupled (band-pass) resonant circuits

5.3.1 Methods of coupling

Radio signals carrying intelligence occupy a band of frequencies and circuits must be able to accept the whole of that band whilst rejecting all others. When tuned circuits are coupled together in the correct manner they form such a band-pass circuit. The coupling may be either by mutual inductance between the two inductances of the circuits as in *Figure 5.3* or through discrete electrical components as in *Figure 5.4*.

Mutual inductance can be defined in terms of the number of flux linkages in the second coil produced by unit current in the first coil. The relationship is:

$$M = \frac{\text{flux linkages in 2nd coil produced by current in 1st coil}}{\text{current in 1st coil}} \times 10^{-8}$$

82

Figure 5.3 Mutual inductance (low impedance) coupling

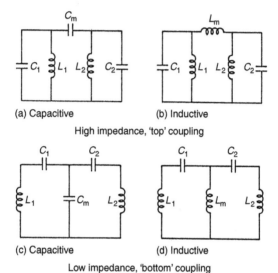

(a) Capacitive   (b) Inductive

High impedance, 'top' coupling

(c) Capacitive   (d) Inductive

Low impedance, 'bottom' coupling

Figure 5.4 Electrical coupling

where M = mutual inductance in henries. The e.m.f. induced in the secondary is $e_2 = -j\omega M I_1$ where I_1 is the primary current.

The maximum value of mutual inductance that can exist is $\sqrt{L_1 L_2}$ and the ratio of the actual mutual inductance to the maximum is the coefficient of coupling, k:

$$k = \frac{M}{\sqrt{L_1 L_2}}$$

The maximum value of k is 1 and circuits with a k of 0.5 or greater are said to be close coupled. Loose coupling refers to a k of less than 0.5.

An advantage of coupling using discrete components is that the coupling coefficient is more easily determined. Approximations for k for the coupling methods shown in *Figure 5.4* are:

(a) $k = \dfrac{C_m}{\sqrt{C_1 C_2}}$ where C_m is much smaller than $(C_1 C_2)$

(b) $k = \dfrac{\sqrt{L_1 L_2}}{L_m}$ where L_m is much larger than $(L_1 L_2)$

(c) $k = \dfrac{\sqrt{C_1 C_2}}{C_m}$ where C_m is much larger than $(C_1 C_2)$

(d) $k = \dfrac{L_m}{\sqrt{L_1 L_2}}$ where L_m is much smaller than $(L_1 L_2)$

5.3.2 Response of coupled circuits

When two circuits tuned to the same frequency are coupled their mutual response curve takes the forms shown in *Figure 5.5*, the actual shape depending on the degree of coupling.

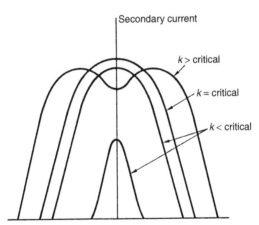

Figure 5.5 Effect of degree of coupling

When coupling is very loose the frequency response and the current in the primary circuit are very similar to that of the primary circuit alone. Under these conditions the secondary current is small and the secondary response curve approximates to the product of the responses of both circuits considered separately.

As coupling is increased the frequency response curves for both circuits widen and the secondary current increases. The degree of coupling where the secondary current attains its maximum possible value is called the critical coupling. At this point the curve of the primary circuit shows two peaks, and at higher coupling factors the secondary response also shows two peaks. The peaks become more prominent and further apart as coupling is increased, and the current at the centre frequency decreases.

References

Langford-Smith, F. (1955). *Radio Designer's Handbook*, Iliffe and Sons, London.
Terman, F.E. (1943). *Radio Engineers' Handbook*. McGraw-Hill, London.

6 Oscillators

6.1 Oscillator requirements

Oscillators generate the frequencies used in radio and electronic equipment. The performance of those which determine the operating frequencies of radio systems is tremendously important. Most oscillators must:

- Generate a precise frequency of high purity.
- Be highly stable, i.e. produce an output constant in frequency and level despite changes in temperature, supply voltage and load.
- Be tunable in frequency.
- Produce minimum noise and microphony (fluctuations in frequency with vibration).

These requirements conflict. A readily tunable oscillator cannot be precise and highly stable, and compromises must be made; either a less stringent specification must be accepted where permissible or the facility for tuning restricted.

Not all oscillators need to produce a pure output, devoid of spurious frequencies. The clock generators in digital circuitry, for instance, produce square waves, but a radio transmitter carrier generator and receiver local oscillator must produce a pure sine wave output if spurious radiations and receiver responses are to be avoided.

6.2 Tunable oscillators

An oscillator is an amplifier with a portion of the output fed back to the input. The feedback must be positive, i.e. in phase with the input, and the loop gain, input back to input via the feedback loop, must be sufficient to overcome the losses in the circuit.

Most radio frequency oscillators – and some audio ones – use inductance and capacitance (LC) tuned circuits as the frequency determining elements. *Figure 6.1* shows two commonly used basic circuits, the Hartley and the Colpitts.

The frequency is determined by the values of L and C_1 (the combined values of C_1 in the Colpitts circuit) and the amount of feedback by the collector choke and C_2. Such circuits produce a very pure output but, principally because the physical dimensions of the frequency

85

(a) Hartley circuit

(b) Colpitts circuit

Figure 6.1 Hartley and Colpitts oscillators

determining components change with temperature, the accuracy of the set frequency is doubtful and is not very stable. Temperature compensation can be applied by selecting a capacitor for C_1 with the correct negative temperature coefficient (assuming that the inductance increases with temperature), inserting temperature compensation for the rise in collector current with temperature and stabilizing the supply voltage. In the design of equipment an oscillator should be built into an area of low temperature change.

6.3 Quartz crystal oscillators

The problems of frequency accuracy and stability are largely overcome by using a quartz crystal as the frequency determining element (see Chapter 7).

Figure 6.2(a) shows a circuit for an oscillator using the crystal's parallel resonant mode. In this circuit, the rising voltage developed across R_e on switch-on is applied via C_1 to the base accelerating the rise of current through the transistor. When saturation is reached the voltage across R_e becomes static and the voltage on the base falls, reducing the transistor current. The oscillations are only sustained

(a) Parallel resonant

(b) Series resonant

Figure 6.2 Quartz crystal oscillators

at the parallel resonant frequency of the crystal where it presents a high impedance between base and collector. C_t enables the parallel resonance of the crystal to be adjusted to a precise frequency.

Figure 6.2(b) shows a series resonant crystal oscillator and here L_t is the tuning inductor of a Colpitts oscillator. The loop gain is adjusted so that the circuit will oscillate only at the series resonant frequency of the crystal where it presents a very low resistance. At other frequencies the crystal presents an increasing impedance in series with L_t, shunted by R which can be of a low value. When first setting the oscillator, L_t is adjusted, with the crystal short-circuited, for oscillation at a frequency close to that desired. The short-circuit is then removed and L_t used as a fine trimmer.

The same circuit will operate at the parallel resonant frequency of the crystal by making C_1 equal to the crystal load capacitance.

The maximum frequency error permitted by the British Radiocommunications Agency specification MPT 1326 for mobile radio equipment designed for 12.5 kHz channel separation in the band 100–300 MHz is plus or minus 1.5 MHz. This is an overall accuracy of 0.0005% over the temperature range $-10\,^{\circ}\text{C}$ to $+55\,^{\circ}\text{C}$. Well-designed standard crystal oscillators meet this specification, but higher stability can be obtained by operating the crystal in an oven at a constant higher temperature.

Until recently equipment which was required to change operating frequency quickly was fitted with several crystals, one for each operating frequency, and a change of frequency was made by selecting the appropriate crystal. Frequency synthesizer circuits are now normally used for such applications.

6.3.1 Overtone oscillators

Piezoelectric crystals can oscillate at more than one frequency. The oscillations of a crystal slab are in the form of *bulk acoustic waves* (BAWs), and can occur at any frequency that produces an odd half-wavelength of the crystal's physical dimensions (e.g. $1\lambda/2, 3\lambda/2, 5\lambda/2, 7\lambda/2, 9\lambda/2$, where the fundamental mode is $1\lambda/2$). Note that these frequencies are not exact harmonics of the fundamental mode, but are actually valid oscillation modes for the crystal slab. The frequencies fall close to, but not directly on, some of the harmonics of the fundamental (which probably accounts for the confusion). The overtone frequency will be marked on the crystal, rather than the fundamental (it is rare to find fundamental mode crystals above 20 MHz or so, because their thinness makes them more likely to fracture at low values of power dissipation).

The problem to solve in an overtone oscillator is encouraging oscillation on the correct overtone, while squelching oscillations at the fundamental and undesired overtones. Crystal manufacturers can help with correct methods, but there is still a responsibility on the part of the oscillator designer. It is generally the case that overtone oscillators will contain at least one $L-C$ tuned circuit in the crystal network to force oscillations at the right frequency.

6.4 Frequency synthesizers

Frequency synthesizers offer the stability of a quartz crystal oscillator combined with the facility to change operating frequency rapidly. They are essential for equipment operating on trunked or cellular networks where the frequency of the mobiles is changed very rapidly on instructions from the network.

6.4.1 Voltage controlled oscillators

The advent of the variable capacitance diode (varicap), where the capacitance across the diode varies according to the applied DC voltage, made the frequency synthesizer a practicality.

When a varicap diode replaces the tuning capacitor in an oscillator the circuit becomes a DC-voltage-controlled oscillator (VCO). The two varicaps in *Figure 6.3* are used to minimize harmonic production and to obtain a greater capacitance change per volt.

Figure 6.3 Voltage variable capacitance diode

The VCO is the circuit that directly generates the output frequency of a frequency synthesizer but, by itself, inherits the problem of frequency stability. To overcome this, the frequency and phase of the VCO are compared with those of a crystal-controlled high stability oscillator. Any frequency or phase difference between the two oscillators creates a DC voltage of the correct sense to change the frequency of the VCO to agree with that of the crystal oscillator.

While the stability of the crystal oscillator is transferred to the synthesizer output, additional noise is produced close to the operational frequency and the elimination of microphony requires careful physical design.

6.4.2 Phase-locked loops

Figure 6.4(a) and *(b)* are diagrams of a simple phase-locked loop (PLL). The outputs of both the crystal oscillator and the voltage-controlled oscillator are fed to the phase comparator which produces pulses whenever there is a frequency or phase difference between the two inputs. The pulses will be either positive or negative depending on the sense of the difference, and their width is dependent upon the magnitude of the differences. The pulses are then fed to a low pass loop filter which smooths them. If the time constant of the filter is sufficiently long it will completely remove the pulses and produce a DC output proportionate to the input pulse width which is applied to the VCO in the right sense to correct the frequency error. To enable the pulses to swing the VCO frequency in either direction, a small bias voltage of about 4 volts is applied to the varicap.

Identical frequencies have been selected for both oscillators in *Figure 6.4* but in practice this is seldom the case. More frequently, the

(a)

(b)

Figure 6.4 Simple phase-locked loop

VCO runs at a higher frequency than the crystal oscillator and a divider is used to equate the frequencies applied to the phase comparator (*Figure 6.5*). Changing the division ratio provides a convenient means of tuning the oscillator.

(a)

(b)

Figure 6.5 Frequency variable phase-locked loop

While the division ratio is 1000, the VCO will run at 100 MHz but if the division ratio is changed to 999 the comparator will produce pulses which, when converted to a DC voltage by the loop filter, will change the frequency of the VCO to 99.9 MHz, and the loop will lock at this new frequency.

The design of the loop filter is critical. Too long a time constant lengthens the settling time when changing frequency, yet if it is too short any deliberate frequency modulation will be removed. In practice, a relatively long time constant is chosen which is shortened by a 'speed up' circuit introduced whenever a channel change is called for.

The above values would enable a radio operating on a system with a channel separation of 100 kHz to change channel, but mobile radio channel separations are 25 kHz, 12.5 kHz or even 6.25 kHz at frequencies from 50 MHz to at least 900 MHz. To change channel at these frequencies a synthesizer must use a high division ratio. With a reference frequency applied to the comparator of 6.25 kHz and an operating frequency of, say, 450 MHz, the frequency select divider must have a ratio of 72 000, and be programmable in steps of 1 with a minimum operating speed of at least 900 MHz. A problem is then that the technology capable of meeting these requirements, emitter coupled logic (ECL), is power hungry, and the preferred LSI low power technology, CMOS, has a maximum operating speed of about 30 MHz. A simple ECL pre-scaler to bring the VCO frequency to about 30 MHz needs a ratio of 20 (500 MHz to 25 MHz). However, every change of 1 in the CMOS divider ratio then changes the total division ratio by 20. The solution is to use a dual modulus pre-scaler.

6.4.3 Dual modulus pre-scaler

The division ratio of the dual modulus pre-scaler (*Figure 6.6*) is programmable between two consecutive numbers, e.g. 50 and 51 (*P* and *P* + 1) and, in conjunction with two CMOS dividers, ÷A and ÷N, provides a fully programmable divider.

Figure 6.6 Programmable frequency synthesizer

The A and N dividers are pre-loaded counters. These count down and when the count value reaches zero they produce an output which changes the division ratio of the pre-scaler. The total division ratio, N_t, is decided by the initial programmed contents of the A and N counters and the setting of the pre-scaler. The initial content of the A counter must be less than that of the N counter.

Consider the pre-scaler set to divide by $P + 1$. For every count of $P + 1$, the contents of the A and N counters are reduced by 1 until the contents of the A counter are zero. The difference between the original contents of the A and N counters, $N - A$, remains in the N counter, and the total count, N_t, up to now, is $A(P + 1)$. At this point the division ratio of the pre-scaler is changed to P. Now, for every P count, the contents of the N counter are reduced by 1 until zero is reached. Under these conditions the total division ratio is given by:

$$N_t = A(P + 1) + (N - A)P$$
$$= AP + A + NP - AP$$
$$= NP + A$$

For example, let $P = 50$ so $P + 1 = 51$, let $N = 10$ and $A = 7$. Then:

$$N_t = 10 \times 50 + 7 = 507$$

Now, change A to 6:

$$N_t = 10 \times 50 + 6 = 506$$

a change of N_t by 1.

Programming a divider

Example:

$$\text{VCO frequency} = 455.6\,\text{MHz}$$

$$\text{Reference frequency} = 12.5\,\text{kHz}$$

Calculate N_t, and the numbers which must be programmed into the A and N counters, assuming $P = 50$:

1. Calculate $N_t = 455.6\,\text{MHz}/12.5\,\text{kHz} = 36\,448$.
2. Divide N_t by P: $36\,488/50 = 728.96$. Make $N = 728$.
3. For A, multiply fraction by P: $0.96 \times 50 = 48$.
4. Check $N_t = NP + A = 728 \times 50 + 48 = 36\,448$.
   Change A to 47: $NP + A = 728 \times 50 + 47 = 36\,447$.
   $36\,447 \times 12.5\,\text{kHz} = 455.5875\,\text{MHz}$, the adjacent 12.5 kHz channel.

6.4.4 Direct digital synthesis

A method of direct digital frequency synthesis replaces the voltage-controlled oscillator by a numerically controlled oscillator (NCO) where the function of the VCO is digitally synthesized.

The direct digital synthesizer generates an analogue sine wave from digital sine wave samples applied to a digital to analogue (D/A) converter. There are limitations to the method in terms of bandwidth and spectral purity.

6.5 Caesium and rubidium frequency standards

Where extra high stability is required for, say, laboratory standards or in quasi-synchronous wide area coverage systems, oscillators utilizing the atomic resonances of substances like caesium and rubidium, although expensive, may be employed.

Caesium oscillators are used to provide standard frequencies such as 1, 5 and 10 MHz with accuracies of $\pm 7 \times 10^{-12}$ over a temperature range of 0 to 50 °C with a long-term stability of 2×10^{-12}.

Rubidium oscillators are used to provide secondary standards and in some quasi-synchronous radio systems. Their accuracy is less than that of caesium, the long-term drift being of the order of 1×10^{-11} per month.

References

Belcher, R. *et al.* (1989). *Newnes Mobile Radio Servicing Handbook*. Butterworth-Heinemann, Oxford.

Winder, S.W. (2001). *Newnes Telecommunications Pocket Book*. Butterworth-Heinemann, Oxford.

7 Piezo-electric devices

7.1 Piezo-electric effect

When electrical stress is applied to one axis of a quartz crystal it exhibits the piezo-electric effect: a mechanical deflection occurs perpendicular to the electric field. Equally, a crystal will produce an e.m.f. across the electrical axis if mechanical stress is applied to the mechanical axis. If the stress is alternating – the movement of the diaphragm of a crystal microphone is an example – the e.m.f. produced will be alternating at the frequency of the movement. If the stress alternates at a frequency close to the mechanical resonance of the crystal as determined by its dimensions, then large amplitude vibrations result. Polycrystalline ceramics possess similar qualities.

Quartz crystals used for radio applications are slices cut from a large, artificially grown crystal. The slices are then ground to the appropriate size to vibrate at a desired frequency. The performance of an individual slice – the crystal as the end user knows it – depends upon the angle at which it was cut from the parent crystal.

Each crystal slice will resonate at several frequencies and if the frequency of the stimulus coincides with one of them the output, electrical or mechanical, will be very large.

The vibrations occur in both the longitudinal and shear modes, and at fundamental and harmonic frequencies determined by the crystal dimensions.

Figure 7.1A shows a typical natural quartz crystal. Actual crystals rarely have all of the planes and facets shown. There are three *optical axes* (X, Y and Z) in the crystal used to establish the geometry and locations of various cuts. The actual crystal segments used in RF circuits are sliced out of the main crystal. Some slices are taken along the optical axes, so are called Y-cut, X-cut and Z-cut slabs. Others are taken from various sections, and are given letter designations such as BT, BC, FT, AT and so forth.

7.1.1 Equivalent circuit of a quartz crystal

A quartz crystal behaves similarly to a very high Q tuned circuit and the equivalent circuit of a crystal is shown in *Figure 7.1B*.

C_1 and L_1 are equivalent to the inductance and capacitance of a conventional tuned circuit and R_1 represents the losses in the quartz

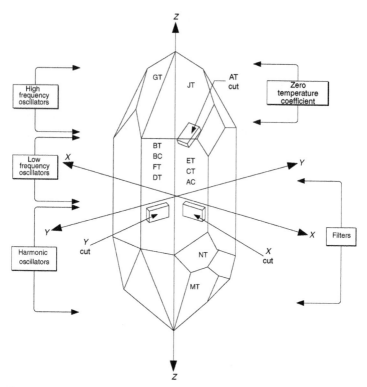

Figure 7.1A Natural quartz crystal

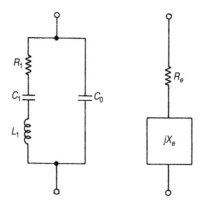

Figure 7.1B Equivalent circuit of a crystal

and the mounting arrangements. C_0, typically 3–15 pfd, represents the shunt capacitance of the electrodes in parallel with the can capacitance. If the oscillatory current is considered, the resonant frequency is decided by the values of C_0 in series with C_1, L_1 and R_1, and all crystals basically resonate in a series mode. *Figure 7.2* illustrates the changes in impedance close to resonance. However when a high impedance, low capacitance, load is connected across the crystal terminals it behaves as a parallel tuned circuit exhibiting a high resistance at the resonant frequency. A crystal operating in the parallel mode oscillates at a higher frequency than that of series resonance.

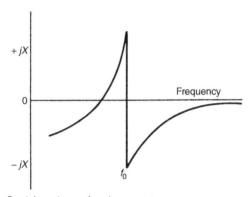

Figure 7.2   Crystal reactance close to resonance

A crystal will resonate at its fundamental frequency or at one or more of its harmonics. As the desired resonant frequency is increased, a crystal slice operating at its fundamental frequency becomes extremely thin and fragile. Consequently, overtone crystals are composed of larger slices of quartz operating close to, but not necessarily at, an exact harmonic of the fundamental frequency. Crystals operating at the 3rd, 5th and 7th harmonics are often employed at frequencies above approximately 25 MHz.

7.2 Quartz crystal characteristics

7.2.1 Resonant frequency

The resonant frequency is determined by the mass of the finished crystal which can be adjusted by grinding and the deposition of gold or other metal onto the crystal faces during manufacture. The adjustment is made to suit the intended method of operation, series or parallel, and at a specific temperature, usually 25°C. When parallel mode is specified, allowance is made for the load or circuit capacitance, usually 20–30 pfd, in parallel with C_0.

7.2.2 Frequency stability

Temperature coefficient

A crystal's resonant frequency varies with temperature and this temperature coefficient is determined by the angle at which the slice was cut from the parent crystal. Commonly used cuts are AT and BT. Because of its better performance AT is the most common.

Typical examples of the temperature coefficients for these are shown in *Figure 7.3*.

The temperature coefficient is specified, usually in parts per million (ppm) per degree C, or as a percentage, over a defined temperature range. The standard European temperature range is −10°C to +60°C. A crystal designed for a restricted temperature range has a better

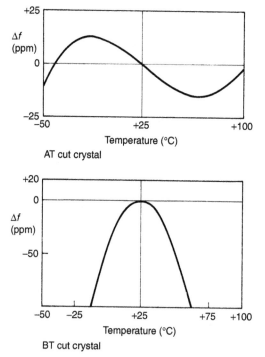

Figure 7.3   Frequency vs. temperature curves AT and BT cut crystals

stability over that range than one designed for operation over a wide temperature range will have when used over a restricted range.

For higher frequency stability crystals may be operated in a temperature-controlled oven operating at a more constant high temperature.

Common frequency tolerance specifications are $\pm0.005\%$ or 0.0025% from $-55°C$ to $+105°C$. These include the frequency errors from all sources, including the calibration tolerance; thus, the temperature coefficient is slightly better than these figures.

Ageing

The resonant frequency shifts with age from that set at production, following a curve similar to that in *Figure 7.4*. Initially the frequency shift for a given period of time is rapid but slows with age. The frequency may shift in either direction, and although it is possible to specify crystals ageing in one direction – high stability oscillators for quasi-synchronous transmission systems is an application – they are selected from a batch, not specifically manufactured. Once a crystal has been operated, a subsequent long period of inactivity can produce a glitch in the ageing curve followed by a higher rate of change for a short time.

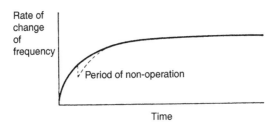

Figure 7.4   Effect of ageing

7.2.3 Load capacitance and pullability

When a crystal is operated in the parallel mode across a low capacitance load the results are a higher frequency and larger output voltage to the load. Increasing the load capacitance causes a reduction in frequency approaching that of series resonance.

The change in frequency that can be achieved by varying the load capacitance – a small trimmer capacitor is often connected across the crystal for this purpose – is the crystal's pullability. A typical pullability is from -1 ppm/pfd to -20 ppm/pfd for a total shunt capacitance of 40 pfd ($C_0 + C_{load}$).

7.2.4 Activity, effective series resistance (ESR) and Q

All these characteristics are interrelated. A crystal's activity, its vibrational response, can be quoted in terms of the effective series resistance. A higher effective series resistance implies lower activity, lower output and lower Q. The usual range of ESRs is from 20 Ω to 100 Ω although higher values occur in some low frequency crystals. Some manufacturers may quote activity levels for crystals for use in a parallel mode as effective parallel resistance (EPR). The EPR is the value of the resistor which, if connected in lieu of the crystal in an oscillator, would give the same output level as the crystal. The higher the EPR, the greater the crystal activity and Q.

7.2.5 Spurious responses

Crystals will resonate at frequencies other than those of the fundamental and harmonic modes for which they were designed; *Figure 7.5* shows the overtone (harmonic) and some typical spurious responses. The spurious responses of overtone crystals can occur with very little separation from the desired overtone frequency requiring very careful oscillator design if they are to be avoided.

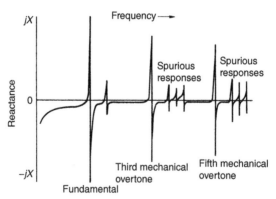

Figure 7.5 Overtone response of a quartz crystal

7.2.6 Case styles

A wide range of mounting styles is available. The American military nomenclature is widely used to describe them and *Figure 7.6* shows the outlines of some of the more popular styles.

Solder seal	HC6/U AA	HC18/U J	–	HC33/U BF	HC25/U CX	MIL IEC
Cold weld	HC36/U DNC	HC43/U DPC	HC45/U EBC	HC47/U AC	HC42/U DQ	MIL IEC
Glass	HC27/U DA	HC26/U CY	– –	– –	HC29/U CZ	MIL IEC
Res weld	HC48/U DNR	HC43/U DPR	HC45/U EBR	HC51/U RF	HC50/U RX	MIL IEC

Solder seal	– –	– –	– –	– –		MIL IEC
Cold weld	HC35/U DKC	– –	– –	– –		MIL IEC
Glass	– –	– –	– –	– –		MIL IEC
Res weld	HC35/U DKR	HC49/4H –	– UM-5	– UM-1		MIL IEC

Figure 7.6 Crystal case styles

7.3 Specifying quartz crystals

The details which must be specified when ordering crystals are:

1. *Frequency.* Normally specified in kHz up to 9999.999 kHz and in MHz from 10.0 MHz upwards except for integer values which are all specified in MHz. The frequency must be described to seven significant figures, otherwise any figure that might follow those given will be taken as zero.
2. *Mounting or holder style.*
3. *Frequency tolerance.* This is the cutting or calibration tolerance acceptable at 25°C. It should be borne in mind that cost rises with increased manufacturing accuracy and a slight adjustment (pullability) is possible in the circuit.

4. *Frequency stability*. Normally specified as a plus or minus value measured over a defined temperature range. A crystal designed for a restricted range has a better performance over that range than one designed for a wider range so it is important not to overspecify.

5. *Temperature range*. The range over which the crystal is required to operate and meet the performance specified in 4. Standard temperature ranges are:

> 0 to 5°C
> −10 to 60°C
> −20 to 70°C
> −30 to 80°C
> −40 to 90°C
> −55 to 105°C
> −55 to 125°C

It is sufficient when ordering from some manufacturers to quote only the lower temperature limit.

For ovened operation the quoted figure, say 80°C, would denote the oven temperature.

6. *Circuit condition*. This specifies the shunt capacitance that the circuit will place across the crystal in parallel mode operation.

7. *Drive level*. The maximum power that the crystal can safely dissipate. 1 mW is a typical value for crystals used in radio transmitters and receivers.

A typical specification therefore reads:

16.66667 MHz	HC49	20	30	10	30
1	2	3	4	5	6

referring to the items listed above

When the crystal is for operation in series mode, it is usually sufficient to replace the last figure with 'S'. The drive level is not normally specified in the ordering details.

7.4 Filters

Both quartz and ceramic materials are used in the production of radio frequency filters. Ceramic filters do not have the same performance as quartz but have the advantages of a lower cost. They are used at lower frequencies and where the higher stability and lower spurious responses of quartz are not essential.

Crystal filters are obtainable at frequencies up to about 45 MHz. Most of these use either a number of discrete crystals arranged in the form of a lattice or a monolithic structure. A single crystal will behave as an extremely narrow band filter and it is possible to use a crystal bar in this way down to a very few kilohertz.

The characteristics of filters can be divided into groups affecting the performance (*Figure 7.7*).

Figure 7.7 Filter characteristics

7.4.1 Passband performance

- *Insertion loss*. The loss at centre frequency, in dB, resulting from the insertion of the filter in a transmission system.
- *Flat loss*. The insertion loss at the frequency of minimum loss within the passband.
- *Attenuation*. The loss of a filter at a given frequency measured in dB.
- *Passband (bandwidth, BW_1)*. The range of frequencies attenuated less than a specified value, typically 3 or 6 dB.
- *Centre frequency (f_0)*. The arithmetic mean of the passband limits.
- *Fractional bandwidth*. A specified frequency, typically the minimum loss point or F_0, from which all attenuation measurements are made.
- *Ripple*. The amplitude difference, in dB, between the maximum peak and minimum passband valley. Both the peak and the valley are defined by a surrounding change in slope, i.e. sign of the amplitude response. This is very important as a high ripple, particularly between a peak and the adjacent trough, produces rapid phase changes as the signal moves across the

passband resulting in audio distortion and corruption in digital signals.

7.4.2 Stopband performance

- *Attenuation*. The output of a filter at a given frequency relative to the defined insertion loss reference.
- *Stopband*. The range of frequencies attenuated by a greater amount than some specified minimum level of attenuation.
- *Transition band (bandwidth, BW_2)*. The range of frequencies differently attenuated between the passband and stopband limits.
- *Shape factor*. The ratio of the bandwidth at some point within the transition region, typically 60 dB, to the specified passband bandwidth. It is given by:

$$\text{Shape factor} \quad \frac{BW_2}{BW_1}$$

- *Spurious attenuation*. The specified minimum level of attenuation received by all non-harmonic related resonances of each crystal resonator within the filter network.

7.4.3 Time domain performance

- *Insertion phase*. The phase shift at the output load (measured at the reference frequency) resulting from the insertion of the filter.
- *Differential phase*. The measurement of phase at a given frequency relative to the phase at the reference frequency.
- *Phase linearity*. The phase error in degrees between the phase points and a straight line drawn through the phase points.
- *Group delay*. The time by which a signal will be delayed before it appears at the filter output, i.e. the derivative of phase with respect to frequency.
- *Differential delay*. The measurement of delay at a given frequency relative to the reference frequency.

7.4.4 Source and load impedance

- *Source impedance*. The impedance of the circuit driving the filter, measured at the reference frequency.
- *Load impedance*. The impedance of the circuit terminating the filter at its output, measured at the reference frequency.

7.4.5 Non-linear effects

- *Maximum input level.* The driving point power, voltage or current level above which intolerable signal distortion or damage to the device will result.
- *Drive level stability.* The ability of the filter to return within a specified tolerance of its original insertion loss, at a specified drive level, after experiencing changing environmental and/or drive level conditions.
- *Drive level linearity.* The maximum permissible variation in insertion loss, per dB change in drive level, measured over a specific dynamic range.
- *Inband intermodulation distortion.* The attenuation, in dB, of third and higher order signal products, inband, relative to the power level of two signals placed within the passband.
- *Out-of-band intermodulation distortion.* The attenuation, in dB, of third and higher order signal products, inband, relative to the power level of two signals placed in the stopband, or one signal in the transition region and the other in the stopband.

A manufacturer's specifications for two stock 10.7 MHz filters are given in *Table 7.1.*

Table 7.1 Manufacturers' specifications for two stock 10.7 MHz filters

Centre freq.	Passband width (plus/minus)	Attenuation bandwidth	Ripple (max)	Ins. loss (max)	Term. impedance ΩS/pfd
10.7 MHz	3.75 kHz (6 dB)	8.75 kHz (45 dB)	2.0 dB	3.0 dB	1.5k/1
		12.5 kHz (60 dB)			
10.7 MHz	7.5 kHz (6 dB)	15.0 kHz (60 dB)	2.0 dB	4.0 dB	3k/1
		20.0 kHz (80 dB)			

7.5 SAW filters and resonators

The piezo-electric effect that some ceramic materials such as lithium niobate exhibit enables useful applications, such as pressure sensors and audible alarms. When piezo-electric material has an electric field applied across it, usually through metal electrodes, the material changes shape. This often takes the form of local material expansion or contraction (depending on the polarity of the applied voltage). In the case of the audible alarm, the material expansion and contraction modulates the air pressure to give an audible tone if the electric field is alternating at audio frequencies.

One form of the piezo-electric effect is the SAW (surface-acoustic-wave) phenomenon, in which signals travel due to an acoustic wave in the ceramic material. To see this effect the piezo-electric material has one side bonded to a metal plate. The upper surface has metal electrodes applied to it, to provide an electric energizing force. The surface wave is created when a material is forced to expand (by the piezo-electric effect) beneath the metal electrodes. Physical bonds between molecules then force adjacent material to expand and this expansion propagates through the material.

The piezo-electric effect is bi-directional, so that if the material between electrodes changes shape, an electric field is generated between the electrodes. This is the effect seen when a piezo-electric gas lighter is used: a sudden force across a piezo-electric element generates high voltages that force a spark to be produced. In the SAW device, the acoustic wave travels along the material and deforms the piezo-electric material between a second set of electrodes, which are located at the other end of the piezo-electric material. A voltage is then formed between these electrodes. By having two sets of electrodes, known as transducers, at opposite ends of a piece of piezo-electric material, a signal can be transmitted through the material by applying a voltage at one end and detecting it at the other.

The range of frequencies that propagate through the piezo-electric material can be controlled by suitable transmit and receive transducer spacing. The propagation frequencies can be further controlled by applying additional metallized areas (described later) between the transmit and receive transducers. Thus SAW devices could be used to make compact and low-cost filters.

SAW resonators use piezo-electric material that is free to vibrate in one direction. The speed of wave propagation and the dimensions of the material are such that the wave reflects back and forth, resonating at a certain frequency. This can replace quartz crystals in many oscillator circuits, where the frequency accuracy is not critical.

7.5.1 Fundamentals of SAW transversal filters

In its simplest form, a transversal SAW filter consists of two transducers with inter-digital arrays of thin metal electrodes deposited on a piezoelectric substrate, such as quartz or lithium niobate (see *Figure 7.8*). The electrodes that comprise these arrays are arranged to have alternate polarities, so that an RF signal voltage of the proper

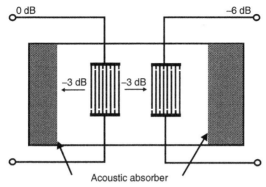

Figure 7.8 Simple transversal SAW filter configuration

frequency applied across them causes the surface of the crystal to alternately expand and contract along its length. This generates the Rayleigh wave, or surface wave, as it is more commonly called.

These inter-digital electrodes are generally spaced at half or quarter wavelength of the operating centre frequency. Since the surface wave or acoustic velocity is 10^5 slower than the speed of light, an acoustic wavelength is much smaller than its free space electromagnetic counterpart. For example, a CW signal at 1000 MHz with a free space wavelength of 300 mm would have a corresponding acoustic wavelength of about 3 μm. This wavelength compressing effect results in the SAW filter's unique ability to provide considerable signal processing or delay in a very small volume.

The wavelength compressing effect of the piezo-electric material has the effect of producing physical limitations at very high and low frequencies. At very high frequencies the electrodes become too narrow to fabricate with standard photo-lithographic techniques. At low frequencies the devices become impracticably large. Consequently, SAW devices are most typically used over the frequency range 10 MHz to about 3 GHz.

The basic SAW transducer is a bi-directional radiator. That is, half of the power (or −3 dB) is directed towards the output transducer while the other half is radiated towards the end of the crystal and is lost. By reciprocity, only half of the intercepted acoustic energy at the output is reconverted to electrical energy; hence, the inherent 6 dB loss associated with this structure (refer to *Figure 7.8*). Numerous second-order effects, such as coupling efficiency, resistive losses, and impedance mismatch, raise the insertion loss of practical filters to about 15 to 30 dB.

A new low-loss structure is the single-phase unidirectional transducer (SPUDT). Unlike the traditional three-phase unidirectional filters, the SPUDT devices are generally as straightforward to fabricate as ordinary bi-directional transducers. In addition, the required impedance matching network is greatly simplified, usually consisting of an L-C network on each port. Most SPUDT structures contain acoustic reflectors within the inter-digital pattern, illustrated schematically in *Figure 7.9* as wider electrodes. These internal reflectors serve to redirect most of the acoustic energy that is normally lost in a conventional bi-directional device. Second-order effects tend to limit the practical insertion loss to about 5 to 12 dB.

Figure 7.9 Single-phase unidirectional filter configuration

7.5.2 SAW device design

A better understanding of SAW device operation can be obtained by understanding how they are designed. SAW filters use finite impulse response design techniques, which are very similar to those used for digital filters. Hence, the Fourier transform is used to relate the time and frequency responses of the transducers and resultant filter. The desired total frequency response characteristics are used to find the impulse responses for the two transducers. These two impulse response shapes are then etched onto the surface of a metallized piezo-electric substrate to form transducers. When an impulse is applied to the transducer, the output wave shape is the same as the shape of the metallized transducer pattern.

To illustrate this relationship, the following examples may be valuable. First, consider the case of the uniform-overlap (or unweighted) transducer as shown in *Figure 7.10*. Since all fingers have the same

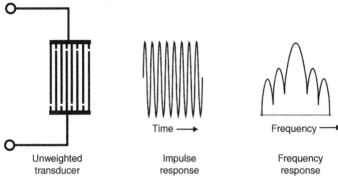

Figure 7.10 Uniform-overlap transducer

length, the impulse response envelope of this transducer is a rectangle. The Fourier transform of a rectangle is a $\sin(x)/x$ function with a 4 dB bandwidth roughly equal to the reciprocal of its time length. This illustrates the important trade-off between bandwidth and size, the narrower the bandwidth: the longer the transducer.

The $\sin(x)/x$ response has 13 dB first side-lobe attenuation, which results in 26 dB of rejection when two of these unweighted transducers are cascaded on the substrate. The 26 dB frequency rejection and poor skirt selectivity make the unweighted transducer suitable for delay lines, but unsuitable for most filter applications where more rejection is needed close to the passband edge.

According to Fourier transform theory, to obtain the rectangular frequency response, a transducer must be built that is infinitely long with electrode lengths that have $\sin(x)/x$ weighting. This is impossible, but the principle behind this example will always hold true: the steeper the skirts (the more rectangular the response) for a given bandwidth, the longer the device must be. The longest possible SAW device would exhibit a narrow bandwidth with a very steep roll-off. The transducer pattern must be limited in length but a gradual tailing off to zero length of the outer electrodes provides close to ideal performance (akin to the windowing function in digital FIR filters).

References

Crystal Product Data Book (1993). International Quartz Devices Ltd, Crewkerne, Somerset.

Winder, S.W. (2001). *Newnes Telecommunications Pocket Book*. Butterworth-Heinemann, Oxford.

8 Bandwidth requirements and modulation

8.1 Bandwidth of signals at base band

8.1.1 Analogue signals

The amount of information and the speed at which it is transmitted determines the base bandwidth occupied by a signal. For analogue signals, the base bandwidth is the range of frequencies contained in the signal; it is not the same as that occupied by a radio frequency carrier modulated by the signal. Examples of base bandwidths are given in *Table 8.1*.

Table 8.1 Base bandwidths

Application	Frequency range (Hz)
Speech	
High fidelity reproduction	15–15 000
Good fidelity	150–7000
Public address	200–5000
Restricted bass and treble	500–4000
Toll quality (good quality telephone line)	300–3400
Communications quality (radio communication)	300–3000
Mobile radio (12.5 kHz channel separation)	300–2700
Music (for FM broadcasting)	30–15 000
Video	60 Hz–4.2 MHz

8.1.2 Digital signals

Bit rate (b/s) and baud rate are terms used to specify the speed of transmitting digital information. Where the duration of all the signalling elements is identical the terms are synonymous, but not where the duration of the information bits differs.

As the term implies, the bit rate is the number of bits transmitted per second but the baud rate (after J.M.E. Baudot, the code's inventor) is the reciprocal of the length, in seconds, of the shortest duration signalling element. *Figure 8.1(a)* shows a binary code pattern where

Figure 8.1 Bit rate and baud rate

all the bits are of equal duration, in this case 1 millisecond; the bit rate is 1000 per second and the baud rate is $1/0.001 = 1000$ also.

On telegraphy systems all the bits may not be of the same duration and *Figure 8.1(b)* shows the pattern for the letter 'J' in the CCITT-2 code as used for teletype. In this code a letter is composed of 5 elements, each of 22 ms duration, but each letter is preceded by a space of 22 ms and followed by a mark of 31 ms. The duration of each character is 163 ms – the time for 7.5 elements – but is comprised of only 7 bits.

The baud rate is $1/0.022 = 45.5$ baud.

The average bit duration is $163/7 = 23.29$ ms.

The average bit rate is $1/0.023 = 43.5$ bits per second.

If the bandwidth were under consideration the baud rate, being faster, would be the figure to use.

A stream of binary coded information is composed of pulses where, say, a pulse (mark) represents digit 1 and the absence of a pulse (space) represents digit 0. The highest frequency contained in the information is determined by the bit – or baud – rate. Because a series of 1s or 0s may be consecutive in the data stream, the pulse repetition rate will vary throughout the message although the bit rate will be constant, and over a long period of time as many spaces will be sent as pulses. Transmitting either a stream of spaces or of marks requires no bandwidth; it is only when a change of state occurs that frequencies are produced. In the duration of one cycle (*Figure 8.1(a)*) two bits may be carried and the maximum *fundamental* frequency contained in the wave is one-half the number of bits per second, i.e. the channel capacity, bits/second, equals twice the bandwidth in hertz.

8.1.3 Channel capacity – Hartley – Shannon theorem

Channel capacity as stated by Hartley's law is, in the absence of noise:

$$C = 2\delta f \log_2 N$$

where
  C = channel capacity, bits per second
  δf = channel bandwidth, Hz
  N = number of coding levels (2 in binary system)

When noise is present, the channel capacity calculated according to the Hartley–Shannon theorem is:

$$C = \delta f \log_2(1 + S/N)$$

where S/N = the ratio of total signal power to total noise power at the receiver input within the bandwidth, δf.

8.2 Modulation

For radio transmission, the low frequency information signal is carried on a radio frequency wave and it must change (modulate) that carrier. The modulation may change the amplitude, frequency or phase of the carrier. Modulation aims to achieve:

1. the transfer of information with the minimum distortion or corruption
2. the modulation of the carrier with the minimum loss of power
3. efficient use of the frequency spectrum.

8.3 Analogue modulation

8.3.1 Amplitude modulation (AM)

There are a number of methods of modulation where the amplitude of the carrier is varied by the information signal but the most commonly used is double sideband AM (DSB). *Figure 8.2* shows a radio frequency carrier modulated by a low frequency signal.

The amount or depth of modulation is expressed as percentage ratio, $m\%$, of the maximum to minimum amplitude:

Mod. depth = $m\%$

$$= \frac{\text{max. amplitude} - \text{min. amplitude}}{\text{max. amplitude} + \text{min. amplitude}} \times 100\%$$

When the modulation is increased to the point where the minimum amplitude falls to zero, 100% modulation occurs. Any further increase in modulation produces spurious, out-of-band frequencies (AM splash), a source of interference for other radio users. For this reason, the depth of amplitude modulation is usually limited to 70%.

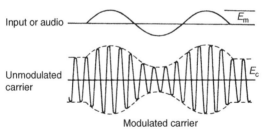

Figure 8.2 Amplitude modulation

An alternative expression for modulation depth is in terms of a modulation index from 0 to 1. The peak carrier voltage in *Figure 8.2* is E_c and the peak modulation voltage, E_m. The modulation index, m, is:

$$m = \frac{E_m}{E_c}$$

Amplitude modulation produces a band of frequencies above and below the carrier frequency – the upper and lower sidebands. The width of each sideband is equal to the highest modulating frequency so the bandwidth of an AM wave is $2 \times$ the highest modulating frequency. To conserve spectrum, the range of modulating frequencies

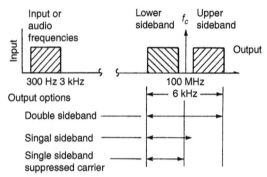

Figure 8.3 Amplitude modulation – sidebands

is restricted. For example, radio communication quality speech is limited to 300 Hz to 3000 Hz. The bandwidth occupied by a double sideband, amplitude modulated carrier for this service is 6 KHz (*Figure 8.3*).

Power relationships in an AM wave

The total power in an AM wave is the sum of powers of the carrier, the upper sideband and lower sideband:

$$P_t = \frac{E_{\mathrm{carr}}^2}{R} + \frac{E_{\mathrm{lsb}}^2}{R} + \frac{E_{\mathrm{usb}}^2}{R}$$

where all values are RMS and R is the resistance in which the power is dissipated. From the peak voltages shown in *Figure 8.2* the power in the unmodulated carrier is:

$$P_c = \frac{E_c/\sqrt{2}^{\,2}}{R} = \frac{E_c^2}{2R}$$

The power in the sidebands

$$P_{\mathrm{lsb}} = P_{\mathrm{usb}} = \left(\frac{mE_c/2}{\sqrt{2}}\right)^2 \div R = \frac{m^2 E_c^2}{8R}$$

$$= \frac{m^2}{4}\frac{E_c^2}{2R}$$

The ratio of the total power in the wave to the carrier power is:

$$\frac{P_t}{P_c} = 1 + \frac{m^2}{2}$$

As m cannot exceed 1, the maximum RMS power in the wave is $P_t = 1.5 P_c$; but if m reaches 1, the peak sum of E_c and E_m is $2E_c$ and so the instantaneous peak power is $2P_c$. Circuitry must be capable of handling this power level without distortion.

Double sideband amplitude modulation wastes power and spectrum. Two-thirds of the power is in the carrier which conveys no information and one sideband is discarded in the receiver. Also, the modulation must be accomplished either in the final power amplifier of the transmitter necessitating a high power modulator, or in an earlier, low power stage when all subsequent amplifiers must operate in a linear, but inefficient, mode.

8.3.2 Double sideband suppressed carrier (DSBSC)

In an amplitude modulated wave the carrier conveys no information yet contains 2/3 of the transmitted power. It is possible to remove the carrier by using a balanced modulator (*Figure 8.4*), and improve the power efficiency by this amount.

C_1 and C_2 are RF bypass capacitors

Figure 8.4 Balanced modulator

In a balanced modulator, the modulating voltage is fed in push–pull to a pair of matched diodes or amplifiers while the carrier is fed to them in parallel. The carrier components in the output cancel leaving the two sidebands. The result is a double sideband suppressed-carrier (DSBSC) wave, which is not sinusoidal, formed by the sum of the two sidebands. The carrier must be re-introduced in the receiver and its accuracy in both frequency and phase is critical.

8.3.3 Single sideband suppressed carrier (SSB or SSBSC)

The advantages of single sideband suppressed carrier transmission over double sideband AM are:

- removal of the carrier saves 2/3 of the total power
- removal of one sideband saves 50% of the remaining power
- an SSBSC transmitter only produces power when modulation is present
- the occupied bandwidth is halved; a spectrum saving
- the received signal-to-noise ratio is improved by 9 dB for a 100% modulated carrier. Halving the bandwidth accounts for 3 dB, the remainder from the improved sideband power to total power ratio. The S/N ratio improves further with lower modulation levels
- reduced susceptibility to selective fading and consequent distortion.

Two methods of generating a single sideband wave are in general use. One filters out the unwanted sideband after removal of the carrier by a balanced modulator. The other is a phase shift method (*Figure 8.5*). Here, the modulating signal is fed to two balanced modulators with a 90° phase difference. The output from both modulators contains only the sidebands but, while both upper sidebands lead the input carrier voltage by 90°, one of the lower sidebands leads it by 90° and the other lags it by 90°. When applied to the adder, the lower sidebands cancel each other while the upper sidebands add.

Figure 8.5 Phase shift production of single sideband

A single sideband AM wave modulated by a sinusoid consists of a constant amplitude signal whose frequency varies with the frequency of the modulating wave. Note that this is not the same as FM: the frequency in SSB does not swing to either side of the carrier. It is higher than the carrier frequency if the upper sideband is transmitted, and lower if the lower sideband is selected. The single sideband waveform is sinusoidal and, although the frequency of the reintroduced carrier must be highly accurate ($\pm 2\,Hz$), the phase is unimportant making a single sideband receiver less complex than one for DSBSC.

On some systems a pilot carrier is transmitted and the transmitter output power is then specified in terms of peak envelope power (pep), the power contained in a wave of amplitude equal to the pilot carrier and transmitted sideband power. Where no pilot carrier is transmitted, the power is specified as peak sideband power (psp).

8.3.4 Frequency modulation (FM)

Both frequency and phase modulation (both may be referred to as angle modulation) effectively vary the frequency of the carrier rather than its amplitude. Frequency modulation varies the carrier frequency directly but its amplitude remains constant regardless of the modulating voltage. Angle modulation is employed at VHF and above for both communications and broadcasting services.

When frequency modulated, a carrier frequency either increases or decreases when the modulation voltage is positive and varies in the opposite sense when the modulating voltage is negative (*Figure 8.6*). The amount of modulation, i.e. the 'deviation' of the carrier from its nominal frequency, is proportional to the amplitude, and not the frequency, of the modulating voltage. The modulation index, M, is defined as the deviation divided by the modulating frequency:

$$M = \frac{f_d}{f_m} \text{ or } \frac{\omega_d}{\omega_m}$$

where
f_d = deviation in hertz
f_m = modulating frequency in hertz

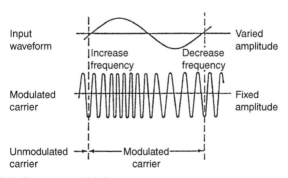

Figure 8.6 Frequency modulation

$\omega_d = 2\pi f_d$ = deviation in radians

$\omega_m = 2\pi f_m$ = modulating frequency in radians

The 'deviation ratio', D, is given by:

$$D = \frac{f_{d(max)}}{f_{m(max)}}$$

The peak deviation is the maximum amount of modulation occurring (equivalent to 100% amplitude modulation). With FM, this is limited only by the need to conserve spectrum; there is no technical limit where distortion occurs, as with 100% AM. The maximum permitted deviation for a service is determined by regulation.

The bandwidth of a frequency modulated signal is made up of the carrier and a series of sidebands, sometimes referred to as sidecurrents, spaced apart from each other at the modulating frequency. The number of sidebands is proportional to the modulation index and their amplitudes decrease with spacing from the carrier. It is generally considered satisfactory to transmit those sidebands $M + 1$ in number, with amplitudes greater than 10% of that of the carrier for that modulation index (*Figure 8.7*). The sidebands occur on both sides of the carrier and:

$$1st\ order\ sidebands = f_c \pm f_m$$

$$2nd\ order\ sidebands = f_c \pm 2f_m,\ etc.$$

When the modulation index approaches 6 a good approximation of the bandwidth (δf) required for an FM transmission is $2(f_d + f_m)$ Hz. For example, speech 300–3000 Hz, max deviation 15 kHz (relevant to a VHF, 50 kHz channel spacing system):

$$\delta f = 2(f_d + f_m) = 2(15 + 3)\,kHz = 36\,kHz$$

$$M = 15/3 = 5$$

The bandwidth is also given by $\delta f = f_m \times$ highest needed sideband $\times$ 2. From Bessel functions (*Figure 8.7*), a modulation index of 5 requires the 6th order sideband to be transmitted ($M + 1$). Therefore:

$$\delta f = f_m \times 6 \times 2 = 3 \times 6 \times 2 = 36\,kHz$$

For specific values of M the carrier of an FM wave disappears. The successive disappearances and the modulation index are given in *Table 8.2*.

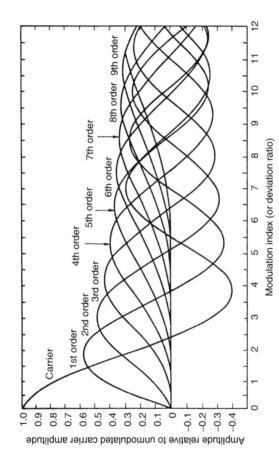

Figure 8.7 Amplitudes of components of FM wave with variation of modulation index

Table 8.2 Successive disappearances and modulation index

Order of disappearance	Modulation index M
1	2.40
2	5.52
3	8.65
4	11.79
5	14.93
6	18.07
$n(n > 6)$	$18.07 + \pi(n - 6)$

Systems where the modulation index exceeds $\pi/2$ are considered to be wide band FM (WBFM), those with a modulation index lower than $\pi/2$, narrow band FM (NBFM). The bandwidth of an NBFM signal is $2f_{m(max)}$.

Pressure on spectrum necessitates narrower channel spacings for communications and, at VHF, 12.5 kHz is normal with ±2.5 kHz as the maximum permitted deviation. Lower standards of performance with a restricted modulation index and a highest modulating frequency of 3 kHz have had to be accepted.

For transmitters used on 12.5 kHz channel spaced systems the highest modulating frequency is, in practical terms, 2700 Hz because the specification requires the frequency response to fall above 2.55 kHz.

The modulation index on such systems, assuming a highest modulating frequency of 3 kHz, is $2.5/3 = 0.8333$ ($<\pi/2$ and so system is NBFM) and the amplitude of the 2nd order sideband is $<10\%$ of carrier amplitude (*Figure 8.7*), so the bandwidth = $2f_{m(max)} = 6$ kHz.

8.3.5 Phase modulation

The end result of phase modulation is frequency modulation, but the method of achieving it and the definition of the modulation index is different. Phase modulation is used in VHF and UHF transmitters where the carrier frequency is generated directly by a crystal oscillator. The frequency of a crystal oscillator can be varied by only a few radians but if the oscillator frequency is multiplied to produce the final carrier frequency the phase variation is also multiplied to produce a frequency deviation.

In frequency modulation deviation is proportional to the modulating voltage, but in phase modulation the frequency deviation is proportional to both the modulating voltage and frequency.

The phase modulation, ϕ_d radians, equals the modulation index so, for phase modulation, the modulation index is $\phi_d = \omega_d/\omega_m$.

Phase modulation has, therefore, a frequency response for the deviation that rises at 6 dB per octave of the modulating frequency. The flat frequency response of FM can be produced on a phase modulated transmitter by installing a filter with a response falling at 6 dB per octave in the audio amplifier, and the rising response of PM can be produced in an FM transmitter with a rising response filter.

8.3.6 Pre- and de-emphasis

Phase modulation, and FM modified to give a rising frequency response (pre-emphasis), offer an improved signal-to-noise ratio in the receiver. The +6 dB per octave response produced in the transmitter is restored to a flat response in the receiver by a −6 dB per octave filter in the audio circuitry which reduces both the enhanced levels of the higher speech frequencies and the high frequency noise (de-emphasis).

8.3.7 Merits of amplitude and frequency modulation

The advantages and disadvantages of AM and FM are given in *Table 8.3*.

Table 8.3 Advantages and disadvantages of AM and FM

	Advantages	Disadvantages
AM	Simple modulators and de-modulators	Susceptible to man-made noise
	Narrower bandwidth than wide-band FM	Audio strength falls with decreasing RF signal strength
		Inefficient power usage
		Limited dynamic range
		Transmitter output power not easily adjusted
FM	Less susceptible to noise	Wider bandwidth
	Constant audio level to almost the end of radio range	
	Capture effect in receiver	Capture effect may be undesirable, e.g. aviation communications
	More power-efficient	
	Transmitter output power easily adjustable	

8.3.8 Stereo FM radio

An important advantage of FM over AM radio broadcasting is the availability of stereo sound. This requires the transmission and

122

reception of the left and right audio signals, to produce a stereo image. However, the introduction of stereo had to be achieved while still catering for large numbers of listeners who wanted to continue receiving in mono. The introduction of stereo FM radio had to go unnoticed by anyone using old mono radios. Without this condition, a simpler method could have been chosen.

The standard FM stereo system used by broadcasters around the world is illustrated in the block diagram of *Figure 8.8*. It uses frequency division multiplexing (FDM) to combine the two signals of the left and right channels. The signals are filtered to limit the bandwidth to 15 kHz. The left (L) and right (R) signals are then added to produce a sum signal and subtracted one from the other to produce a difference signal.

Figure 8.8   FM stereo transmitter

The sum signal provides a monophonic signal, which provides a baseband signal for the frequency modulator. This was the technique used in mono FM and thus was the obvious choice for stereo FM, to allow backward compatibility. An mono FM radio can receive this signal and recover the combined L and R channels, thereby satisfying the requirement for providing unchanged service to mono radios.

The difference signal is used to amplitude modulate a 38 kHz sinewave. By utilizing a balanced mixer, double sideband suppressed carrier (DSBSC) is generated. However, the modulation method must take into account the ease of demodulation. In particular, demodulating a DSBSC signal can be difficult. Both frequency and phase of the carrier are needed to perform faithful demodulation.

In the stereo system the DSBSC demodulation problem is dealt with by including a 19 kHz pilot tone in the broadcast. This tone is generated by a divide-by-two frequency converter circuit, which takes the 38 kHz carrier and produces the 19 kHz pilot tone. The 19 kHz pilot tone falls midway in the spectral region between the mono sum signal (up to 15 kHz) and below the DSBSC difference signal information. The DSBSC signal extends from 23 kHz to 53 kHz, since the input modulating signals are band limited to 15 kHz. The DSBSC output is added to the baseband (L and R sum) signal and the 19 kHz pilot tone before being sent to the FM modulator.

A mono FM receiver ignores the stereo information by using a filter after its FM demodulator to block everything above 15 kHz. It passes the combined L and R channel signal, which is monophonic.

A stereo receiver has an additional circuit after the FM demodulator to detect and demodulate the DSBSC signal. The stereo receiver detects a 19 kHz pilot tone and uses this to generate a 38 kHz signal. This is then used to demodulate the DSBSC signal that carries the L and R channel difference information. The stereo receiver then has both the sum and difference signals, which is all that is needed to recreate the separate left and right signals. Separation is achieved by adding and subtracting sum and difference signals.

The noise power spectral density of a demodulated FM signal tends to increase with the square of the modulation frequency. This is why pre-emphasis is used to boost the high frequency baseband signals for maintaining the signal-to-noise ratio of the transmitted signal. However, this means that there will be more noise in the 23 kHz to 53 kHz band used for the difference signal than for the 0–15 kHz band used for the sum signal. Consequently a significantly higher input signal level is required to receive a stereo transmission compared with a mono signal for the same output signal-to-noise ratio. Thus stereo reception requires far higher radio signal levels than for mono reception and is more susceptible to interference from other radio sources.

8.4 Digital modulation

8.4.1 Data processing

Filtering

A data pulse with a sharp rise- and fall-time produces harmonic frequencies and requires a wide bandwidth if it is to maintain its

124

shape during transmission. Consequently, to transmit the data over a limited bandwidth the pulses must be shaped to reduce the harmonic content as much as possible without impairing the intelligibility of the signal. This is accomplished by the use of low-pass (Gaussian) filters of which the result is a string of smoother pulses, often referred to as 'tamed' (*Figure 8.9(b)*). Tamed FM permits high data rates within a limited channel bandwidth whilst maintaining acceptable adjacent channel interference levels.

(a) Unprocessed 'raw' data

(b) Filtered 'tamed' data

(c) Manchester encoded, 'tamed' data

Figure 8.9   Data processing

Manchester encoding

A serious problem with the transmission of binary data is that unless the clocks in the transmitter and receiver are synchronous the digits become confused, particularly where a continuous string of 1s or 0s occurs. Manchester encoding makes 1s change state from 1 to 0, and 0s from 0 to 1 during each digit period (*Figure 8.9(c)*), facilitating the synchronization of the clocks and rendering the digits more easily recognizable.

Gray coding

For binary data the number of signalling levels is not restricted to two (mark and space). Multi-level signalling has the advantage that one signalling element carries the information for more than one information bit, thus reducing the bandwidth requirement.

If the number of levels is increased to m, where $m = 2^n$ (i.e. $n = \log_2 m$), the m-level data symbol is represented by n binary digits of 0 or 1 (see quaternary phase shift keying, Section 8.4.5). For example, in a quaternary data signal, $m = 4 = 2^2$ giving the binary sequences 00, 01, 10, 11 for each signalling level. This process is Gray coding where there is only one digit difference for each transition between adjacent levels. The number of levels is not restricted to four and 16 level quadrature amplitude modulation (QAM), where $n = 4$, is an efficient system.

In a Gray coded signal where each element contains n bits ($n = \log_2 m$) and the signalling element rate is B bauds (elements per second), the transmission rate is $B \log_2 m$ bits per second.

8.4.2 On/off and amplitude shift keying

On/off keying

The earliest modulation method. A continuous radio frequency wave (CW) is interrupted in a recognizable pattern (Morse code). To provide audibility the carrier is heterodyned with a beat frequency oscillator (BFO) in the receiver. The use of a modulated continuous wave (MCW) eliminates the need for a BFO but the bandwidth of the signal is increased. The problem with on/off keying is the lack of a reference level. If the signal strength temporarily falls below the sensitivity threshold of the receiver it appears to the operator as a series of spaces.

Binary amplitude shift keying (ASK or BASK)

This shifts the level of an audio frequency subcarrier which then modulates a radio frequency carrier.

Because the level of a subcarrier is changed, AM sidebands are produced. Also, because the keyed waveform is non-sinusoidal harmonics occur. The occupied sub-carrier bandwidth for ASK is:

$$\text{Bandwidth} = 2B$$

where B = bit repetition rate (bits/second).

When the RF carrier is modulated its bandwidth is $2(f_c + B)$ where f_c = subcarrier frequency.

8.4.3 Frequency shift keying (FSK)

Although used for conveying digital information, frequency shift keying in reality employs frequency modulation. In its original form, developed for HF transmission, FSK changes the carrier frequency to indicate a 1 or a 0 but retains the nominal carrier frequency as a reference and to represent a mark. A downwards shift of carrier frequency by 170 Hz represents a space in the HF radio system.

Table 8.4 Standard ITU-T frequencies for various bit rates

Bit rate (bps)	Frequencies		Subcarrier frequency
	0	1	
600	1700	1300	1500
1200	2100	1300	1700
up to 300	1180	980	1080
	1850	1650	1750

Modern FSK uses two different modulation frequencies to represent 1s and 0s. If intersymbol interference (ISI) is to be avoided the separation of the tones must be more than half the bit rate, and a factor of 0.7 is often used. Standard ITU-T frequencies for various bit rates are given in *Table 8.4*.

The base bandwidth requirement is:

$$\text{Bandwidth} = f_2 - f_1 + 2B$$

The bandwidth of a modulated carrier is:

$$\text{RF bandwidth} = 2(f_2 + B)(\text{narrow band FM})$$

Minimum shift keying (MSK) is a form of FSK where the frequency deviation is equal to half the bit rate.

Gaussian minimum phase shift keying (GMSK)

Similar to MSK, the Gaussian filters improve the adjacent channel performance against a small cost (approximately 1%) in ISI while achieving high data transmission rates.

8.4.4 Fast frequency shift keying (FFSK)

Fast frequency shift keying may either amplitude or frequency modulate the carrier. In binary FFSK, the data is changed in a modem to tones of 1800 Hz to represent binary 0 and 1200 Hz to represent binary 1. During transmission a binary 1 consists of 1 cycle of 1200 Hz, f_1, and a 0, $1\frac{1}{2}$ cycles of 1800 Hz, f_2, i.e. a bit rate of 1200 bps. For acceptable intersymbol interference the distance between the tones cannot be less than half the bit rate and the 600 Hz separation in FFSK represents the fastest signalling speed – hence the description – and minimum bandwidth. For this reason it is sometimes called minimum frequency shift keying (MFSK). The base bandwidth is the same as for FSK, but the RF bandwidth depends upon the system deviation. For example, on a 12.5 kHz channel spaced system carrying FFSK and complying with Radiocommunications Agency Code of Practice MPT 1317, deviation = 60% of system deviation = 1.5 kHz:

$$f_1 = 1200 \, \text{Hz}$$

$$f_2 = 1800 \, \text{Hz}$$

$$B = 1200 \, \text{bps}$$

$f_{\text{max}} = f_2 + B = 3 \, \text{kHz}$ and the mod. index $m = \text{max. dev.}/f_{\text{max}} < \pi/2$ so the system is NBFM and the bandwidth $= 2(f_2 + B) = 6 \, \text{kHz}$.

Where more data states than binary are to be transmitted, multistate FSK (M-ary FSK) is also possible where M may be up to 32 states.

Both FFSK and M-ary FSK are well suited to radio transmission as the change of state occurs while the signals are passing through zero, avoiding sudden phase changes (*Figure 8.10(a)*).

The minimum distance between the tones used in FFSK of 0.5 times the bit rate is not ideal for immunity to intersymbol interference (ISI).

Logic 0 three half cycles 1800 Hz
Logic 1 two half cycles 1200 Hz

1200 Hz 1800 Hz 1200 Hz

1/1200 s   1/1200 s   1/1200 s

(a) FFSK modulation

1 0 1 1 0 0 0 1

Message
baseband
signal

PSK

(b) PSK modulation

Figure 8.10   FFSK and PSK

128

A separation of 0.7 times the bit rate would improve the ISI but increase the bandwidth.

8.4.5 Phase shift keying

There are several variants of phase shift keying. Binary phase shift keying (BPSK or PSK) changes the phase of the carrier by 180° at the zero crossing point (*Figure 8.10(b)*). No carrier frequency is present with PSK as half the time the carrier is multiplied by +1 and the other half by −1 and cancels out, but the reference phase of the carrier must be re-inserted at the receiver. The bandwidths occupied are the same as for ASK, i.e.:

$$\text{Baseband} = 2B$$

$$\text{RF bandwidth (AM or NBFM)} = 2(f_c + B)$$

Differential phase shift keying (DPSK) advances the phase 90° or 270° at each change of logic state (*Figure 8.11*). Changing phase only at a

Data	Phase change
0	+90°
1	+270°

(a) DPSK modulation

Dibit	Phase change
00	+45°
01	+135°
11	+225°
10	+315°

(b) QPSK modulation

Figure 8.11   DPSK and QPSK

change of logic state saves bandwidth which, for DPSK, is equal to the bit rate.

An important advantage of both FFSK and PSK over FSK is that because the moment of change is predefined it is possible to recover data more accurately. However, the transition between signalling states is not smooth requiring large and rapid phase shifts. Multilevel systems with less phase shift between elements are preferable.

Quaternary, or quadrature, phase shift keying (QPSK) is a four-level Gray-coded signalling method with $90°$ phase shift between adjacent signalling elements (*Figure 8.11(b)*). If the signal is considered as a vector the points at $\pi/4$ ($45°$), $3\pi/4$ ($135°$), $5\pi/4$ ($225°$) and $7\pi/4$ ($315°$) represent the transition points between states and the binary data.

8.5 Spread spectrum transmission

The spread spectrum technique spreads the carrier containing the information over a very wide bandwidth, typically 1.25 MHz, using pseudo-noise generation techniques as described in Chapter 12. The transmitter uses what is in effect a digital key to spread the bandwidth and the receiver is equipped with an identical key for despreading. A number of users with different keys can occupy the same band at the same time. The system operates well in poor signal-to-noise or high interference environments.

A continuous wave (CW) transmission concentrates all the radiated energy on a single frequency (*Figure 8.12*). Amplitude modulation and narrow band FM widen the radiated bandwidth, reducing the energy at the carrier frequency and per kHz. Wide band FM carries the process a stage further until with spread spectrum the band width is increased to the extent that the signal almost disappears into the noise floor.

Spreading of the bandwidth is achieved by multiplying the digitally modulated signal by a spreading code at a much higher bit rate (100–1000 times the signal bit rate). This is done by combining the signal with the output of a random code generator running at 2 or 3 orders of magnitude faster than the binary signal rate. *Figure 8.13* is a block diagram of a spread spectrum system. A clock running at the spreading rate R_c is used to drive both the spreading generator and, after frequency division, the data encoder. The carrier is first BPSK modulated by the encoded data and then in a balanced modulator (the spreading correlator) by the high rate code from the spreading generator. The resultant transmitted bits are referred to as chips to distinguish them from data bits. In the receiver, the clock pulses R_c are recovered and used to drive both the despreading generator and decoder.

Figure 8.12 Comparative energy dispersal

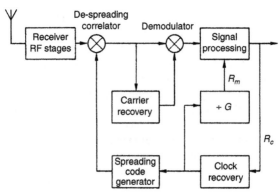

Figure 8.13 CDMA (spread spectrum) system

Many users can be accommodated by allocating each a unique spreading code. It is common to use a pseudo-noise (PN) generator to multiply the bit rate and then to modulate the carrier with either FSK or PSK.

Although spread spectrum is a digital system, in quality of signal there are similarities with analogue:

Analogue	Digital
Signal/noise ratio	Energy per bit, E_b/N_o
Intelligibility, signal/ noise + distortion	Bit error rate

References

Belcher, R. *et al.* (1989). *Newnes Mobile Radio Servicing Handbook.* Butterworth-Heinemann, Oxford.

Kennedy, G. (1977). *Electronic Communications Systems*, McGraw-Hill Kogashuka, Tokyo.

Winder, S.W. (2001). *Newnes Telecommunications Pocket Book.* Butterworth-Heinemann, Oxford.

9 Frequency planning

9.1 International and regional planning

The International Telecommunications Union (ITU) administers the planning and regulation of the radio frequency spectrum on a world-wide basis through the World Administrative Radio Conferences.

For planning purposes the world is divided into three regions as shown in *Figure 9.1*. The boundaries are formed by geographical features suited to the purpose such as seas, high mountain ranges or uninhabited remote areas.

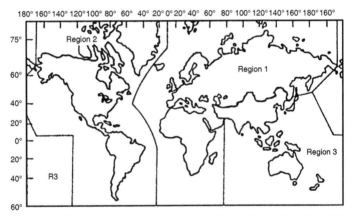

Figure 9.1 ITU defined regions. For purposes of international allocations of frequencies the world has been divided into three regions

The World Administrative Radio Conferences decide the use of blocks of the spectrum, e.g. sound broadcasting, television, marine communications, and the countries permitted to use the blocks for those allotted purposes.

9.2 National planning

Having been allocated blocks of frequencies for a particular type of use, the administration of each country determines the method of allocating frequency bands from each block to the user categories within their country. Every country has a radio regulatory department within its administration. In the UK this is the Radiocommunications Agency of the Department of Trade and Industry.

9.2.1 The role of the Radiocommunications Agency

The aims of the Radiocommunications Agency in the UK, and, in general, the radio regulatory bodies of other countries are:

1. To ensure that the radio frequency spectrum is used in ways which maximize its contribution to national social and economic welfare, having regard to safety of life factors.
2. To make the maximum amount of spectrum available for commercial use.
3. To provide an expert service to government as a whole in the field of radio regulation.

The first stage of national planning is the assignment of the radio frequencies (channels) within a geographic area. Both the allocation of the blocks to users and the geographic assignment of channels if not wisely carried out can result in spectrum pollution – intermodulation products are one source (see Chapter 19) – and unacceptable interference between services and users. The incorrect allocation of blocks may not only cause interference in the home country but, even at VHF and UHF, between adjacent countries. Incorrect assignment of channels causes a more local problem but, depending on the frequencies involved, the undesirable effects can spread over a wide area.

The second stage of national planning is the assignment of discrete channels for use on multi-user sites where the selection of incompatible frequencies causes interference, receiver de-sensitization and, possibly, blocking, and intermodulation products.

Not all channels are allocated directly by the regulatory body. Blocks of channels, usually comprised of two frequencies, may be issued to responsible user organizations: the Home Office for the police and fire services is an example. These organizations then become responsible for the frequency planning and allocation within their user group.

Additional to the allocation of frequencies, the Radiocommunications Agency through the licensing procedure regulates the use of base station sites, the maximum transmitter power, and antenna directivity for each service. It also prepares and publishes technical specifications with which all equipment must comply. A list of the current specifications, of which single copies can be obtained, is available from

The Information and Library Service, Radiocommunications Agency, Wyndham House, 189 Marsh Wall, London E14 9SX.

The Radio Investigation Service (RIS) is the branch of the Agency which, in addition to investigating interference, inspects all radio installations prior to commissioning and, if in order, issues an inspection certificate. No station is permitted to operate without a certificate and may not be modified subsequent to the issue of the certificate.

Current policy throughout Europe is leading towards the de-regulation of radio communications while safeguarding the protection from interference. Allocation of frequencies by pricing is also under consideration on the basis that a scarce resource, the spectrum, will be allocated to the users having the greatest need.

9.3 Designations of radio emissions

Radio emissions should be expressed in a three-symbol code form, which defines the exact nature of carrier, signal and transmitted information. The first symbol defines the carrier, the second symbol defines the signal, and the third symbol defines the information.

First symbol

A	Double-sideband amplitude-modulated
B	Independent sideband amplitude-modulated
C	Vestigial sideband amplitude-modulated
D	Amplitude- and angle-modulated simultaneously, or in a predefined sequence
F	Frequency modulated
G	Phase modulated
H	Single-sideband, full carrier
J	Single-sideband, suppressed carrier
K	Amplitude-modulated pulse sequence
L	Width-modulated pulse sequence
M	Position phase modulated pulse sequence
N	Unmodulated carrier
P	Unmodulated pulse sequence
Q	Pulse sequence in which carrier is angle-modulated during the pulse period
R	Single-sideband, reduced or variable level carrier
V	Pulse sequence with a combination of carrier modulations, or produced by other means
W	Carrier is modulated by two or more of angle, amplitude, and pulse modes, simultaneously or in a defined sequence
X	Other cases

Second symbol

0	No modulating signal
1	Digital signal without modulating sub-carrier
2	Digital signal with modulating sub-carrier
3	Analogue signal
7	Two or more channels with digital signals
8	Two or more channels with analogue signals
9	Composite system with one or more channels of digital signals and one or more channels of analogue signals
X	Other cases

Third symbol

A	Aural telegraph
B	Automatic telegraph
C	Facsimile
D	Data
E	Telephony (and sound broadcasting)
F	Television
N	No information transmitted
W	Combination of any of the above
X	Other cases

9.4 Bandwidth and frequency designations

A four symbol code should be used to express bandwidth and frequency to three significant figures. A letter to denote the unit of frequency is placed in the position of the decimal point, where the letters and bandwidths are:

Letter	Bandwidth
H	Below 1000 Hz
K	Between 1 and 999 kHz
M	Between 1 and 999 MHz
G	Between 1 and 999 GHz

So, a frequency of 120 Hz is 120H, while a frequency of 12 Hz is 12H0 etc.

9.5 General frequency allocations

VLF, LF, MF (frequency in kHz)

10.0	140.5	Fixed; maritime; navigation
140.5	283.5	Broadcast

255.0	526.5	Radio navigation; fixed
526.0	1606.5	Broadcast
1606.5	1800.0	Maritime and land mobile; fixed
1810.0	1850.0	Amateur (shared in UK)
1850	2000	Amateur
1850	2045	Fixed; mobile
2045	2173.5	Maritime mobile; fixed
2160	2170	Radiolocation
2173.5	2190.5	Mobile
2190.5	2194	Maritime
2194	2625	Fixed; mobile
2300	2498	Broadcast
2625	2650	Maritime mobile
2650	2850	Fixed; mobile
2850	3155	Aero mobile

HF (frequency in kHz)

3155	3400	Fixed; mobile
3200	3400	Broadcast
3400	3500	Aero mobile
3500	3800	Amateur, fixed; mobile
3800	4000	Amateur (region 2 only)
3800	3900	Fixed; mobile
3800	3950	Aero mobile
3950	4000	Fixed; broadcast
4000	4063	Fixed; maritime mobile
4063	4438	Maritime mobile
4438	4650	Fixed; mobile
4650	4750	Aero mobile
4750	5060	Fixed; mobile; broadcast
5060	5480	Fixed; mobile
5450	5730	Aero mobile
5730	5950	Fixed; mobile
5950	6200	Broadcast
6200	6525	Maritime mobile
6525	6765	Aero mobile
6765	7000	Fixed; mobile
7000	7100	Amateur
7100	7300	Amateur (region 2 only)
7100	7300	Broadcast (regions 1 and 3)
7300	8195	Fixed
8100	8815	Maritime mobile
8815	9040	Aero mobile

9040	9500	Fixed
9500	10 000	Broadcast
10 000	10 100	Aero mobile
10 100	11 175	Fixed
10 100	10 150	Amateur
11 175	11 400	Aero mobile
11 400	11 650	Fixed
11 650	12 050	Broadcast
12 050	12 230	Fixed
12 230	13 200	Maritime mobile
13 200	13 360	Aero mobile
13 360	13 600	Fixed
13 600	13 800	Broadcast
13 800	14 000	Fixed
14 000	14 350	Amateur
14 350	15 000	Fixed
15 000	15 100	Aero mobile
15 100	15 600	Broadcast
15 600	16 360	Fixed
16 360	17 410	Maritime mobile
17 410	17 550	Fixed
17 550	17 900	Broadcast
17 900	18 030	Aero mobile
18 030	18 068	Fixed
18 068	18 168	Amateur
18 168	18 780	Fixed
18 780	18 900	Maritime mobile
18 900	19 680	Fixed
19 680	19 800	Maritime mobile
19 800	21 000	Fixed
21 000	21 450	Amateur
21 450	21 850	Broadcast
21 850	21 870	Fixed
21 870	22 000	Aero mobile
22 000	22 855	Maritime mobile
22 855	23 200	Fixed; mobile
23 200	23 350	Aero mobile
23 350	24 890	Fixed; mobile
24 890	24 990	Amateur
25 010	25 070	Fixed; mobile
25 070	25 210	Maritime mobile
21 210	25 550	Fixed; mobile

25 550	25 670	Radio astronomy
25 670	26 100	Broadcast
26 100	26 175	Maritime mobile
26 175	28 000	Fixed; mobile
28 000	29 700	Amateur
29 700	30 000	Fixed; mobile

VHF, UHF (frequencies in MHz)

30.0	50.0	Fixed; mobile
47.0	68.0	Broadcast (TV)
50.0	52.0	Amateur (UK)
50.0	54.0	Amateur (regions 2 and 3)
68.0	74.8	Fixed; mobile
70.0	70.5	Amateur (UK)
74.8	75.2	Aero navigation
75.2	87.5	Fixed; mobile
87.5	108	Broadcast (FM)
108	118	Aero navigation
118	137	Aero mobile
137	138	Spacecraft; satellites
138	144	Aero mobile; space research
144	146	Amateur
146	148	Amateur (regions 2 and 3 only)
146	174	Fixed; mobile
156	174	Maritime mobile
174	230	Broadcast (TV)
220	225	Amateur (USA)
230	328.6	Fixed; mobile
328.6	335.4	Aero navigation
335.4	400	Fixed; mobile
400	410	Space research; meteorology
410	430	Fixed; mobile
430	440	Amateur; radiolocation
440	470	Fixed; mobile
470	855	Broadcast (TV)
855	1300	Fixed; mobile
902	928	Amateur (USA)
934	935	Citizens band (UK)
1240	1325	Amateur
1300	1350	Aero navigation
1350	1400	Fixed; mobile
1400	1429	Space (uplink); fixed
1429	1525	Fixed; mobile

1525	1600	Space (downlink)
1600	1670	Space (uplink)
1670	1710	Space (downlink)
1710	2290	Fixed; mobile
2290	2300	Space (downlink); fixed
2300	2450	Amateur; fixed
2310	2450	Amateur (UK)
2300	2500	Fixed; mobile
2500	2700	Fixed; space (downlink)
2700	3300	Radar
3300	3400	Radiolocation; amateur
3400	3600	Fixed; space (uplink)
3600	4200	Fixed; space (downlink)
4200	4400	Aero navigation
4400	4500	Fixed; mobile
4500	4800	Fixed; space (downlink)
4800	5000	Fixed; mobile
5000	5850	Radio navigation; radar
5650	5850	Amateur
5850	7250	Fixed; space (uplink)
7250	7900	Fixed; space (downlink)
7900	8500	Fixed; mobile; space
8500	10 500	Radar; navigation
10 000	10 500	Amateur
10 700	12 700	Space (downlink); fixed
12 700	15 400	Space (uplink); fixed
17 700	20 000	Space (up/down); fixed
24 000	24 250	Amateur

9.6 Classes of radio stations

AL   Aeronautical radionavigation land station
AM   Aeronautical radionavigation mobile station
AT   Amateur station
AX   Aeronautical fixed station
BC   Broadcasting station, sound
BT   Broadcasting station, television
CA   Cargo ship
CO   Station open to official correspondence exclusively
CP   Station open to public correspondence
CR   Station open to limited public correspondence
CV   Station open exclusively to correspondence of a private
      agency

DR	Directive antenna provided with a reflector
EA	Space station in the amateur-satellite service
EB	Space station in the broadcasting-satellite service (sound broadcasting)
EC	Space station in the fixed-satellite service
ED	Space telecommand space station
EE	Space station in the standard frequency-satellite service
EF	Space station in the radiodetermination-satellite service
EG	Space station in the maritime mobile-satellite service
EH	Space research space station
EJ	Space station in the aeronautical mobile-satellite service
EK	Space tracking space station
EM	Meteorological-satellite space station
EN	Radionavigation-satellite space station
EO	Space station in the aeronautical radionavigational-satellite service
EQ	Space station in the maritime radionavigation-satellite service
ER	Space telemetering space station
ES	Station in the intersatellite service
EU	Space station in the land mobile-satellite service
EV	Space station in the broadcasting-satellite service (television)
EW	Space station in the earth exploration-satellite service
EX	Experimental station
EY	Space station in the time signal-satellite service
FA	Aeronautical station
FB	Base station
FC	Coast station
FL	Land station
FP	Port station
FR	Receiving station only, connected with the general network of telecommunication channels
FS	Land station established solely for the safety of life
FX	Fixed station
GS	Station on board a warship or a military or naval aircraft
LR	Radiolocation land station
MA	Aircraft station
ME	Space station
ML	Land mobile station
MO	Mobile station
MR	Radiolocation mobile station

MS	Ship station
ND	Non-directional antenna
NL	Maritime radionavigation land station
OD	Oceanographic data station
OE	Oceanographic data interrogating station
OT	Station open exclusively to operational traffic of the service concerned
PA	Passenger ship
RA	Radio astronomy station
RC	Non-directional radio beacon
RD	Directional radio beacon
RG	Radio direction-finding station
RM	Maritime radionavigation mobile station
RT	Revolving radio beacon
SM	Meteorological aids station
SS	Standard frequency and time signal station
TA	Space operation earth station in the amateur-satellite service
TB	Fixed earth station in the aeronautical mobile-satellite service
TC	Earth station in the fixed-satellite service
TD	Space telecommand earth station
TE	Transmitting earth station
TF	Fixed earth station in the radiodetermination-satellite service
TG	Mobile earth station in the maritime mobile-satellite service
TH	Earth station in the space research service
TI	Earth station in the maritime mobile-satellite service at a specified fixed point
TJ	Mobile earth station in the aeronautical mobile-satellite service
TK	Space tracking earth station
TL	Mobile earth station in the radiodetermination-satellite service
TM	Earth station in the meteorological-satellite service
TN	Earth station in the radionavigation-satellite service
TO	Mobile earth station in the aeronautical radionavigation-satellite service
TP	Receiving earth station
TQ	Mobile earth station in the maritime radionavigation-satellite service
TR	Space telemetering earth station
TS	Television, sound channel
TT	Earth station in the space operation service

TU   Mobile earth station in the land mobile-satellite service
TV   Television, vision channel
TW   Earth station in the earth exploration-satellite service
TX   Fixed earth station in the maritime radionavigation-satellite
       service
TY   Fixed earth station in the land mobile-satellite service
TZ   Fixed earth station in the aeronautical
       radionavigation-satellite service

9.7 Radio wavebands

Frequency band	Frequency	Wavelength	Waveband definition
VLF	3 to 30 kHz	1 00 000 to 10 000 m	myriametric
LF	30 to 300 kHz	10 000 to 1000 m	kilometric
MF	300 to 3000 kHz	1000 to 100 m	hectometric
HF	3 to 30 MHz	100 to 10 m	decametric
VHF	30 to 300 MHz	10 to 1 m	metric
UHF	300 to 3000 MHz	1 to 0.1 m	decimetric
SHF	3 to 30 GHz	10 to 1 cm	centimetric
EHF	30 to 300 GHz	1 to 0.1 cm	millimetric
EHF	300 to 3000 GHz	0.1 to 0.01 cm	decimillimetric

Reference

Pannell, W.M. (1979). *Frequency Engineering in Mobile Radio Bands.* Granta
Technical Editions, Cambridge.

10 Radio equipment

10.1 Transmitters

10.1.1 Transmitter functions

The functions of all transmitters and the terminology used to describe them, irrespective of the modulation method, are:

1. To generate the radio frequency carrier and amplify it to an appropriate power level; the RF power output.
2. To modulate the carrier with the intelligence to the pre-determined level; the modulation depth for AM, the deviation for FM or PM. The process must introduce the minimum noise and distortion, and prevent the modulation from exceeding the permitted level.
3. Radiate the minimum signals at frequencies outside the permitted bandwidth. Out-of-band or spurious radiation is strictly controlled by the Radiocommunications Agency MPT specifications.

10.1.2 Amplitude-modulated transmitters

Figure 10.1 is a block diagram of an amplitude modulated transmitter; in this case a quartz crystal oscillator generates the carrier frequency, although a frequency synthesizer could equally well be used. The carrier frequency in an AM transmitter is usually generated either at the transmitted frequency or one of its subharmonics; 2nd, 3rd or 6th subharmonic frequencies are typical choices depending on the final frequency.

The output of the oscillator is amplified to the level of the specified power output and if the oscillator runs at a subharmonic a frequency multiplier stage will be included, as in *Figure 10.1*, in the amplifier chain before the final stage, the power amplifier (PA). A tuned filter in the aerial circuit removes from the output unwanted frequencies which might cause interference with other users. A matching circuit correctly matches the impedance of the filter circuit to that of the aerial to ensure maximum power transfer.

In the audio circuits, the speech input from the microphone is processed by controlling the range of frequencies it contains and limiting its amplitude. This eliminates the risk of over-modulation and the production of out-of-band frequencies. Over-modulation produces

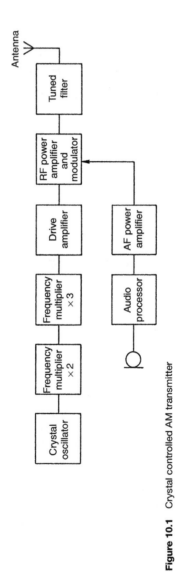

Figure 10.1 Crystal controlled AM transmitter

out-of-band frequencies in any transmitter, but with AM once 100% modulation is exceeded frequencies are produced across a wide range of the spectrum: a disastrous situation for other users. After processing, the audio is amplified in an AM transmitter to a high level and applied to the RF power amplifier to vary its output in the form illustrated in *Figure 8.2*.

AM transmitters are power inefficient. The RF PA cannot operate in class C. It must be linear so as not to distort the speech, and the audio, when it is the power amplifier which is modulated, must be amplified to a high power. If the modulation is applied at a lower power level (in an early stage of the amplifier chain) all the subsequent amplifiers must operate in a linear, but inefficient, manner.

The power output of an AM transmitter is normally specified in terms of the RMS value of the carrier power but the average and peak powers will depend on the depth of modulation, 100% modulation producing a peak power of twice the carrier peak power (see Chapter 8).

10.1.3 Angle modulated transmitters

Figure 10.2 is a block diagram of a frequency modulated transmitter using a frequency synthesizer for carrier generation. The frequency is generated at the final frequency and RF amplifiers raise the power level to that specified for the transmitter output. When a crystal oscillator is used for carrier generation, it must operate at a very low frequency because a quartz crystal oscillator can be frequency modulated by a few radians only. Several stages of the RF amplifier chain then operate as frequency multipliers. Similar filter and aerial matching circuits to those of an AM transmitter are necessary in the output arrangements of the transmitter.

The audio processing circuitry is similar to that for AM transmitters but the modulation (deviation) is applied to a stage operating at a very low RF power level (directly to the VCO in a synthesizer-equipped transmitter and immediately following the crystal oscillator in a direct crystal controlled one); high power audio is not necessary. An additional simple circuit in an FM transmitter may be included to enhance the higher audio frequencies at a rate of 6 dB/octave. This is pre-emphasis and has the merit of improving the level of speech to noise at the receiver (signal-to-noise ratio). The audio level must still be limited because, while the effect of over-deviation is not as disastrous as over-modulation in an AM transmitter, increasing deviation produces a steadily increasing range of frequencies, known as side-currents, outside the permitted bandwidth.

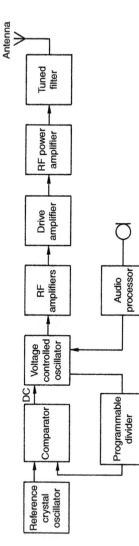

Figure 10.2 Synthesizer controlled FM transmitter

Angle modulated transmitters are more power efficient than amplitude modulated because the modulation is applied at a low power level and also, as no audio frequencies are directly present in the RF amplifier and PA stages, these can operate efficiently in class C.

Phase modulated (PM) transmitters, because the phase shift with modulation is very small, generate the carrier at a very low frequency and use direct crystal control. The frequency is then multiplied many times, thirty-two to thirty-six is common, up to the final frequency. After multiplication, the phase modulation which was originally a few radians has effectively become frequency modulation. The phase modulation process produces pre-emphasis inherently. As far as the user is concerned there is no practical difference between phase modulation and frequency modulation with added pre-emphasis.

10.1.4 Transmitter specifications

In the UK, the Radiocommunications Agency issues specifications with which all equipment must comply. Other countries have their own regulatory bodies, such as the FCC in the USA. These specifications are concerned principally with the prevention of interference and obtaining the maximum use of the frequency spectrum. The characteristics defined in the MPT specifications and other features which affect the user, apart from the physical dimensions, are:

- *Supply voltage.*
- *Operational frequency band.*
- *Modulation method.*
- *Channel separation.*
- *RF power output and impedance.* Output ranges from about 0.5 W to 5 W for hand-portables and 5 W to 25 W for mobiles. The maximum power permitted on a system will be specified in the Licence. Output impedance is commonly 50 Ω.
- *Spurious emissions.* The level of these is critical for the prevention of interference with other users on different frequencies. The limit for VHF and UHF is a maximum of 0.25 μW.
- *Residual noise.* Not always quoted by manufacturers, it is the noise level existing on an unmodulated carrier. A typical figure is better than −40 dB referred to full deviation.
- *Audio frequency distortion.* Typically <3% and usually measured with a modulating frequency of 1 kHz at 60% modulation.
- *Audio frequency response.* This is the variation of modulation level over the audio frequency spectrum. Typically within +1 dB to −3 dB over a frequency range of 300 to 3000 Hz (2.55 kHz for 12.5 kHz

channel spacing equipment). It may be quoted with reference to a pre-emphasis curve.

- *Switching bandwidth.* This is the frequency range over which the transmitter will operate without retuning and without degradation of performance. Much equipment is now specified to cover a complete frequency band, e.g. 146–174 MHz, without retuning.

10.2 Receivers

10.2.1 Receiver functions

A receiver's functions are:

- Detect a weak signal; the minimum level, which may be as low as 0.25 microvolts, defines the receiver sensitivity.
- Amplify a received signal and maintain the information contained in a minimum strength signal at a minimum of 12 dB above the electrical noise level (signal-to-noise ratio). If the audio distortion produced in the receiver is also taken into account the above figure becomes the signal-to-noise + distortion (Sinad) ratio. As the signal is increased, the ultimate Sinad should attain 50 to 55 dB.
- Separate the wanted signal from any unwanted ones which may be very close in frequency (the adjacent channel may be 12.5 kHz away at UHF); the selectivity.
- Recover the information from the carrier; demodulation.
- Amplify the audio information to a level suitable for operating a loudspeaker; the audio power output. The audio amplification must introduce the minimum distortion.
- Disenable the audio amplifiers in the absence of signal to cut out the electrical noise. This is done by the mute or squelch circuit.

10.2.2 Types of receiver

It is possible to amplify directly the incoming RF signal to a level suitable for demodulation. This is done in a tuned radio frequency (TRF) receiver, but these are seldom used today because of the problems of obtaining sufficient selectivity and gain at one radio frequency, and the difficulty of retuning a number of RF stages to change frequency. Almost all receivers designed for analogue communications now operate on the superheterodyne principle where the incoming radio frequency is converted to a lower, more manageable, intermediate frequency (IF). The fixed IF means that only the oscillator and, possibly, one RF amplifier stage need retuning for a change of channel.

The lower frequency of the IF facilitates the acquisition of adequate gain with stability and selectivity.

There is little difference in the layout of receivers for AM and FM except that the circuits perform their functions differently. *Figure 10.3* is a block diagram of a typical FM receiver with a crystal controlled local oscillator.

When two frequencies are applied to a non-linear circuit such as the mixer, they combine to produce other frequencies, their sum and difference being the strongest. The superheterodyne mixes a locally generated frequency with the received signal to produce a, usually lower, frequency retaining the modulation of the received signal. Commonly used values for this intermediate frequency (IF) are 465 kHz for MF and HF receivers and 10.7 MHz for VHF and UHF. At these fixed frequencies the necessary high amplification with low noise and stability and the required selectivity are easier to obtain. The local oscillator (injection) frequency = signal frequency ± IF frequency.

Superheterodyne receivers are susceptible to a particular form of interference. Assume a local oscillator frequency of 149.3 MHz is mixed with a wanted signal of 160 MHz to produce the IF frequency of 10.7 MHz. A signal of 138.6 MHz would also combine with the local oscillator to produce 10.7 MHz. The frequency of this spurious response is the image, or second channel frequency. The IF amplifier cannot discriminate against it so some degree of selectivity must also be provided in the RF amplifier and input circuitry. Intermediate frequencies are chosen which are a compromise between ease of obtaining adjacent channel discrimination and image frequency rejection.

The IF amplifier contains a block filter, either crystal or ceramic (see *Figure 7.7*), necessary to discriminate between channels adjacent in frequency. The design of the filter is crucial. It must be wide enough to accommodate the band of frequencies present in the modulation plus an allowance for frequency drift and its response over this band must be uniform with minimal ripple, particularly if data is to be received, yet its response must be of the order of -100 dB at the frequency of the adjacent channel.

Apart from the demodulator the main difference between AM and FM receivers lies in the operation of the IF amplifier. The IF stages in an AM receiver are linear, although their gain is variable. Part of the IF amplifier output is rectified and used to control the gain to provide automatic gain control (AGC). An increase of signal above a predetermined level, with delayed AGC, causes a reduction in IF gain maintaining a sensibly constant audio output level.

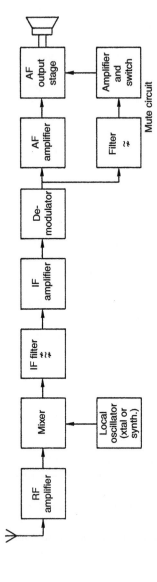

Figure 10.3 FM single superheterodyne receiver

The IF amplifier in a FM receiver possesses a very high gain, some 100 dB, and is non-linear. On receipt of a signal, or even with only the receiver noise, it runs into limitation, cutting off both positive and negative peaks of the signal or noise. This gives FM its constant level audio output over a wide range of signal levels. It also produces the capture effect where a strong signal, fully limiting, completely removes a weaker signal. A signal difference of some 6 dB is required to provide effective capture.

Most receivers employ only one change of frequency, single superheterodynes, but double superheterodynes are occasionally used at VHF and above. A double superheterodyne changes the frequency twice, perhaps to 10.7 MHz for the first IF and then, using a fixed frequency crystal second local oscillator, to a lower, often 1.2 MHz or thereabouts, second IF. The result is greater gain and selectivity but the incorporation of a second oscillator and mixer increases the number of possible spurious responses.

A variation of a very old receiver circuit, the Autodyne, forgotten in about 1914, is now finding favour in receivers for digitally modulated signals. Under its new names of Homodyne or zero-IF receiver it lends itself to the purpose. In this type of receiver the local oscillator runs at the same frequency as the incoming signal, hence the zero-IF. Frequency or phase shifts of the incoming signal representing the data emerge from the mixer at base band and are applied directly to the processing circuits.

In a communications receiver the noise generated in the aerial and RF stages in the absence of a signal is amplified to what may be, when demodulated, an unacceptable level. To eliminate the annoyance the loudspeaker is switched off during no-signal periods by a squelch or mute circuit. The mute circuit rectifies the high frequency noise at the demodulator and uses it to switch off the audio amplifier. When a signal is received the limiting action of the IF amplifier of an FM receiver depresses the noise in favour of the signal. Some AM mobile receivers use additional FM circuitry to provide improved mute action.

10.2.3 Noise figure

An ideal receiver would generate no noise and the signal-to-noise ratio, in a receiver of given bandwidth, would be determined by the level of the signal at the base of the antenna compared with the noise produced in the antenna. The noise factor of the ideal receiver is the number of times the signal power must exceed the antenna noise power to produce a 1:1 ratio at the receiver. It is given by (see Section 1.5.1):

$$\frac{e_{antenna}^2 (\text{e.m.f.})}{4kTBR}$$

When the receiver input impedance is matched to the antenna impedance, half the power is dissipated in the antenna and the noise factor is 2 (3 dB). In a practical receiver, the noise generated in the RF amplifier is the most significant, and the noise figure is the sum of the RF amplifier noise plus all the preceding losses. Figures between 4 and 6 dB are common and the higher the noise figure, the worse the receiver sensitivity.

Signal-to-noise ratio (SNR, S/N or S_N)

Receivers are evaluated for quality on the basis of *signal-to-noise ratio* (S/N or 'SNR'), sometimes denoted S_N. The goal of the designer is to enhance the SNR as much as possible. Ultimately, the minimum signal level detectable at the output of an amplifier or radio receiver is that level which appears just above the noise floor level. Therefore, the lower the system noise floor, the smaller the *minimum allowable signal*.

Noise factor, noise figure and noise temperature

The noise performance of a receiver or amplifier can be defined in three different, but related, ways: *noise factor* (F_N), *noise figure* (NF) and *equivalent noise temperature* (T_E); these properties are definable as a simple ratio, decibel ratio or Kelvin temperature, respectively.

Noise factor (F_N). For components such as resistors, the noise factor is the ratio of the noise produced by a real resistor to the simple thermal noise of an ideal resistor.

The noise factor of a radio receiver (or any system) is the ratio of output noise power (P_{NO}) to input noise power (P_{NI}):

$$F_N = \left[\frac{P_{NO}}{P_{NI}} \right]_{T=290/\text{K}}$$

In order to make comparisons easier the noise factor is usually measured at the standard temperature (T_o) of 290 K (standardized room temperature); although in some countries 299 K or 300 K are commonly used (the differences are negligible). It is also possible to define noise factor F_N in terms of the output and input signal-to-noise ratios:

$$F_N = \frac{S_{NI}}{S_{NO}}$$

where
S_{NI} is the input signal-to-noise ratio
S_{NO} is the output signal-to-noise ratio

Noise figure (NF). The *noise figure* is the frequency used to measure the receiver's 'goodness', i.e. its departure from 'idealness'. Thus, it is a *figure of merit*. The noise figure is the noise factor converted to decibel notation:

$$NF = 10\log(F_N)$$

where
NF is the noise figure in decibels (dB)
F_N is the noise factor
log refers to the system of base-10 logarithms

Noise temperature (T_e). The noise 'temperature' is a means for specifying noise in terms of an equivalent temperature. That is, the noise level that would be produced by a resistor at that temperature (expressed in degrees Kelvin). Evaluating the noise equations shows that the noise power is directly proportional to temperature in degrees Kelvin, and also that noise power collapses to zero at the temperature of Absolute Zero (0 K).

Note that the equivalent noise temperature T_e is *not* the physical temperature of the amplifier, but rather a theoretical construct that is an *equivalent* temperature that produces that amount of noise power in a resistor. The noise temperature is related to the noise factor by:

$$T_e = (F_N - 1)T_0$$

and to noise figure by

$$T_e = KT_0 \log^{-1}\left[\frac{NF}{10}\right] - 1$$

Noise temperature is often specified for receivers and amplifiers in combination with, or in lieu of, the noise figure.

Noise in cascade amplifiers

A noise signal is seen by any amplifier following the noise source as a valid input signal. Each stage in the cascade chain amplifies both signals and noise from previous stages, and also contributes some additional noise of its own. Thus, in a cascade amplifier the final stage sees an input signal that consists of the original signal and noise amplified by each successive stage plus the noise contributed by earlier stages. The overall noise factor for a cascade amplifier can be calculated from *Friis' noise equation*:

$$F_N = F_1 + \frac{F_2 - 1}{G1} + \frac{F_3 - 1}{G1G2} + \cdots$$
$$+ \frac{F_N - 1}{G1G2 \cdots G_{n-1}}$$

where
  F_N is the overall noise factor of N stages in cascade
  F_1 is the noise factor of stage-1
  F_2 is the noise factor of stage-2
  F_N is the noise factor of the nth stage
  $G1$ is the gain of stage-1
  $G2$ is the gain of stage-2
G_{n-1} is the gain of stage $(n - 1)$.

As you can see from Friis' equation, the noise factor of the entire cascade chain is dominated by the noise contribution of the first stage or two. High gain, low noise radio astronomy RF amplifiers typically use low noise amplifier (LNA) circuits for the first stage or two in the cascade chain. Thus, you will find an LNA at the feedpoint of a satellite receiver's dish antenna, and possibly another one at the input of the receiver module itself, but other amplifiers in the chain might be more modest (although their noise contribution cannot be ignored at radio astronomy signal levels).

The matter of signal-to-noise ratio (S/N) is sometimes treated in different ways that each attempts to crank some reality into the process. The signal-plus-noise-to-noise ratio (S + N/N) is found quite often. As the ratios get higher, the S/N and S + N/N converge (only about 0.5 dB difference at ratios as little as 10 dB). Still another variant is the SINAD (signal-plus-noise-plus-distortion-to-noise) ratio. The SINAD measurement takes into account most of the factors that can deteriorate reception.

10.2.4 Signal-to-noise ratio and bandwidth

To obtain the maximum signal-to-noise ratio the bandwidth of every circuit must be designed to admit its operational band of frequencies only. The wider its bandwidth the more noise a circuit admits, and the more the bandwidth exceeds that needed, the worse becomes the signal-to-noise ratio. There is a linear ratio between bandwidth and noise power admitted: doubling the bandwidth doubles the noise power.

Improvements in local oscillator crystal frequency stability has resulted in improved signal-to-noise ratios by reducing the necessary width of the IF filter.

Demodulation, the recovery of the audio from the IF bandwidth affects the signal-to-noise ratio. The relationship is complex, but consider two examples. First, for 12.5 kHz channel spacing FM:

> The audio frequency range is 300 Hz to 3000 Hz
> Bandwidth, $b = 2700$ Hz
> The modulation index $M = f_{d(max)}/f_{m(max)}$
> Signal/noise out $= 3M^2/2b \times$ signal/noise in
> For a 12.5 kHz PMR channel:
> $\quad M = 2500/3000 = 0.83$
> $\quad b = 2.7$ kHz
> $\quad 3M^2/2b = 0.38$
> A 0.38 times reduction in power is -4.8 dB. Demodulation in this case worsens the signal-to-noise ratio by some 5 dB, and if a signal-to-noise ratio of 12 dB is required at the loudspeaker, 17 dB is needed at the input to the demodulator.

When channel separations were 25 kHz, and deviation 5 kHz, the situation was:

> Audio frequency range, 300 Hz to 3000 Hz
> Bandwidth, $b = 2700$ Hz
> For a 50 kHz channel:
> $\quad M = 5000/3000 = 1.66$
> $\quad b = 2.7$ kHz
> $\quad 3M^2/2b = 1.54$
> A 1.54 times gain in power is $+1.8$ dB. In this case the demodulation improved the signal-to-noise ratio slightly. Reducing the bandwidth would have the same effect.

For AM, the signal-to-noise ratio is dependent on the modulation depth. The demodulation process reduces the signal-to-noise ratio by

6 dB but the recovered audio is less than the IF bandwidth by a factor of 3:1 which compensates for the demodulation loss. Reducing the modulation depth degrades the signal-to-noise ratio by 6 dB for every halving of the modulation depth.

10.2.5 Receiver specifications

The important features of receiver specifications are:

- *Sensitivity.* The minimum signal to which a receiver will respond. For an AM receiver, the generally accepted standard is the signal (30% modulated with sinusoidal tone, either 400 Hz or 1 kHz) required to provide an audio output of 50 mW. For an FM receiver, the standard is the unmodulated signal required to produce a 20 dB reduction in noise. A typical figure is $0.25\,\mu V$ (p.d.) for 20 dB quieting. However, sensitivity is often quoted in terms of either the signal-to-noise ratio or Sinad so, in modern parlance, sensitivity and signal-to-noise ratio are sometimes considered to be synonymous.
- *Signal-to-noise ratio (may be quoted as Sinad, signal-to-noise and distortion).* Typically $0.3\,\mu V$ (p.d.) for 12 dB Sinad.
- *Spurious response attenuation.* Typically better than 80 dB.
- *Adjacent channel selectivity.* Better than 65 dB at 12.5 kHz channel spacing.
- *Cross modulation.* The modulation, in the receiver, of a wanted signal by a stronger, unwanted signal. It is usually caused by non-linearity in the receiver RF stages.
- *Blocking and de-sensitization.* The reduction in sensitivity of a receiver due to overloading of the RF stages when a strong signal is applied. A blocked receiver may take an appreciable time to recover.
- *Audio frequency response.* Typically within $+1\,dB$ to $-3\,dB$ of a 6 dB/octave de-emphasis curve from 300 Hz to 3000 Hz (2.55 kHz for 12.5 kHz channel spacing, above which the response falls more rapidly).
- *Audio output.* Typically 3–4 W. For hand-portables, 100–500 mW. (Distortion may also be quoted, typically better than 5%.)
- *Switching bandwidth.* As for transmitter.
- *Duplex separation.* For a receiver/transmitter combination, the minimum separation between the receiving and transmitting frequencies which will permit duplex operation with minimal degradation of the receiver performance.

10.3 Programmable equipment

The current trend is for many of the functions of radio equipment – even down to the control buttons on the front panel – to be software controlled. In one manufacturer's equipment the only screwdriver-adjustable control is that for setting maximum deviation; this is insisted upon by the Radiocommunications Agency for use in the UK.

Software control enables radio sets to be cloned so that once a single piece of equipment is programmed with its frequencies, power output, selective calling, etc., the whole of a fleet of mobiles can be identically programmed in a short time. Some of the functions can be programmed by the user, but not all, and different manufacturers permit different degrees of programming.

The functions are allocated degrees of priority which determine which functions can be changed by each class of person. Again, some of the priorities may be altered by the user but the highest priority functions are installed in a programmable read only memory (PROM) by the manufacturer and can only be altered by replacing the PROM, in most instances a job for the manufacturer.

The essential equipment for major reprogramming is an IBM or equivalent computer and the equipment manufacturer's software which is supplied as a package complete with hardware interface and instructions.

Cloning may be carried out using an inter-connecting cable between the equipment, or restricted reprogramming via a unit supplied by the equipment manufacturer for the purpose.

The functions which may be programmed from a computer are:

- Channel frequencies.
- Transmitter power and deviation.
- Receiver squelch setting (referred to Sinad ratio). Selective calling details including encode or decode only, type of signalling, extension of first tone and call sign.
- Transmission time out timer.
- Channel busy light. Transmitter inhibit on busy channel.
- Alert tones. Tone decoder indicator.
- Low battery indicator. Battery saver.

Restricted reprogramming may include:

- Channel frequencies, including any offsets.
- Lock out of any channels.

- Channel spacing.
- Channel search priority.
- Frequency stability.
- Signalling type.
- Control functions.
- Synthesizer reference crystal (the frequency can be trimmed to compensate for ageing).
- Transmitter power.
- CTCSS (encode and decode).
- Timers.
- Alerts.

Pagers are also programmable. The functions which can be changed include:

- Code number.
- Alert tone/vibrator/repeat.
- Urgency.
- Out of range warning.
- Printer on/off.
- Language of display.

References

Belcher, R. *et al.* (1989). *Newnes Mobile Radio Servicing Handbook.* Butterworth-Heinemann, Oxford.
Langford-Smith, F. (1955). *Radio Designer's Handbook*, Iliffe, London.

11 Microwave communication

11.1 Microwave usage

Microwaves are loosely considered to be those at frequencies between 1 GHz (30 cm) and 100 GHz (0.3 cm). Principal ground communications usage is for point-to-point links carrying information and control signals on systems such as multiplexed communication – including data networks, telemetry and mobile radio. The propagation of the higher frequencies, 30–100 GHz, is being studied for possible future use in micro-cellular radio-telephone systems. These higher frequencies are also used for satellite communications and radar applications. Microwave frequencies up to 3 GHz have now been reserved for mobile use.

11.2 Propagation

The path loss is higher at microwave frequencies than at VHF and UHF: *Figure 1.5* charts the free space loss between isotropic radiators for microwaves at frequencies of 1, 2, 3.5 and 7 GHz over distances of 10 to 1000 km. The free-space loss between isotropic radiators is given by

$$\text{Free space loss, d}B = 32.4 + 20\log_{10} d + 20\log_{10} f$$

where d is in km and f in MHz.

The free-space loss between practical antennas is given by:

$$\text{Loss, dB} = 10\log_{10}\left(\frac{(4\pi d)^2}{\lambda^2}\frac{1}{G_t G_r}\right)$$

where
d = path length, metres
λ = wavelength, metres
G_t = power gain of transmitting antenna
G_r = power gain of receiving antenna

The antenna gains are expressed relative to an isotropic radiator (not in dB).

Absorption varies with atmospheric humidity, and as energy is also absorbed by the ground the path height therefore has an effect on the losses. Absorption by rain is a factor at the higher frequencies but, although considered to be insignificant below about 3 GHz, it has an indirect effect. Wet foliage, for example, produces considerable absorption at frequencies as low as 450 MHz.

Waves of millimetric lengths are special cases and narrow bands of very high absorption due to resonance effects exist at 22 and 183 GHz for water vapour, and 60 and 119 GHz for oxygen. Non-resonant attenuation occurs due to scatter from rain, hail and snow. The attenuation increases with frequency as the wavelength approaches the dimensions of a raindrop. Bands of very low absorption, 'atmospheric windows', where the water vapour and oxygen attenuations are very low occur at 37, 97, 137 and 210 GHz.

Objects close to a path may severely affect the received signal and a proposed path must be examined for the likely effects of these at the planning stage. The effects may be due either to diffraction bending the wave away from the line of sight between the antennas, or reflection causing multi-path signals.

The additional losses in a microwave path caused by objects either intruding into the first Fresnel zone or close to it, and which exhibit 'knife-edge' or 'smooth-sphere' characteristics, are shown in *Figure 11.1*. A negative F/F_1 indicates an intrusion. When the path clearance exceeds 0.6 times the first Fresnel zone radius, the free space loss is achieved.

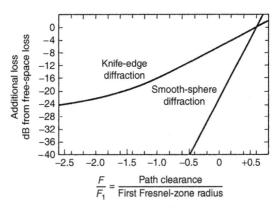

Figure 11.1   Knife-edge and smooth-sphere diffraction

Refraction also affects microwave propagation and temperature inversions may cause ducting resulting in a loss of signal and, possibly, interference. When ducting occurs, a layer of air of low refractive index is formed between two highly refractive layers and a wave may be trapped between them. Under these conditions a signal can be carried an abnormal distance before it can return to Earth resulting in loss of signal at the intended receiver and possibly interference at a distant one on the same frequency.

11.3 *K* factor

The degree and sense of refraction of wave is related to a factor known as K. Refraction normally bends the wave downwards, extending the range, but the refractive index of the atmosphere varies from place to place and with time and height. At times, during an atmospheric inversion for instance, and in some places on the earth the effect is reversed:

When no bending occurs, $K = 1$.
When $K > 1$, the bending is downwards, effectively increasing the Earth's radius.
When $K < 1$, the bending is upwards, effectively reducing the Earth's radius.

For most of the time $K > 1$, 1.33 being accepted as the normal factor, but for small periods K may be less than one.

For a point-to-point link, the path between the transmitting and receiving sites should be a clear line of sight, although by making allowances for refraction, the Earth's radius has, up to now, generally been considered to be effectively increased by the factor of 1.33. However, in the interests of reliability a decrease to 0.7 times the radius (the minimum K factor considered likely to occur) is now often used in link planning. When a link has been planned using a higher K factor, a temporary reduction of K not only reduces the radio horizon but effectively raises objects close to the path, possibly to the point where they become significantly close to the first Fresnel zone.

It is usual when planning a link to plot the profile of a path on paper with curved horizontal graduations to represent the amended Earth's radius. The radio beam can then be drawn as a straight line between the antenna locations. *Figure 11.2* is an example.

11.4 Fresnel zones, reflections and multi-path fading

Signals which arrive at the receiver by more than one path as the results of reflection or diffraction may arrive in any phase relationship

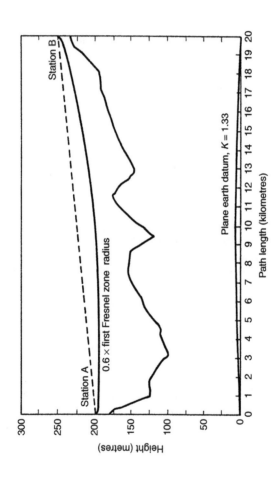

Figure 11.2 Example of path profile

to the direct wave. When they arrive anti-phase to the direct wave, cancellations result. The intensity and phase of the spurious signal may not be constant, thus providing random multi-path fading.

Where a carefully drawn profile of a link path shows there to be a clear line of sight, the effect of waves reflected or diffracted from objects close to the line of the direct wave must then be considered. The effect of these indirect waves can be predicted by calculating where the reflection occurs in relation to a series of ellipsoids which can be drawn around the line-of-sight path between the transmitting and receiving antennas. These ellipsoids, known as the Fresnel zones, contain the points where reflected waves will follow a path of constant length, as shown in *Figure 11.3*.

Figure 11.3 Fresnel zone: reflected path lengths

Waves reflected at the odd-numbered Fresnel zones will travel an odd number of half-wavelengths further than the direct wave but, because a 180° phase change usually occurs in the reflection process, will arrive at the receiver in phase with the direct wave. Waves reflected at even-numbered zones will arrive anti-phase to the direct wave with a cancelling effect. The effect of reflected waves diminishes with reflections from the higher order zones. The radius of a Fresnel zone in metres at the point of intrusion is given by:

First zone, F_1:

$$F_1 = 31.6\sqrt{\frac{\lambda d_1 d_2}{d}}$$

or

$$F_1 = 548\sqrt{\frac{d_1 d_2}{fd}}$$

where
F_1 = radius in metres at point of intrusion
$d_1 + d_2 = d$ (path length in km)
λ = wavelength in metres
f = frequency in MHz

Second zone:

$$F_2 = \sqrt{2} \times F_1$$

Third zone:

$$F_3 = \sqrt{3} \times F_1$$

and so on.

The degree of reflection from an object depends on its nature, the greatest reflection occurring from smooth flat ground or water. Where a path lies over the sea, variations in the path length of a reflected wave due to tides may render a path unusable. When the height of the antenna closest to the sea is varied, the *effect* of the reflected wave passes through a series of minima and maxima and adjustment of the height of that antenna can reduce or, occasionally, overcome the effect.

Atmospheric conditions change giving rise to fading and variations of the multi-path effects. The reliability of a link may be crucial to the success of a complete system and, where a critical path in terms of performance exists, long-term tests are advisable to ensure that variations of propagation do not reduce the reliability to an unacceptable level. Paths which contain obstacles in the line of sight which will cause additional losses are obviously suspect. So are those where objects or large stretches of water or flat ground which might produce diffraction or reflections of the wave lie close to the line of sight.

11.5 Performance criteria for analogue and digital links

The transmission quality for analogue modulated systems is based on the signal-to-noise ratio. The noise is specified relative to a standard test tone level and is commonly expressed as either picowatts psophometrically weighted (pWp) or decibels (dB) of C-message weighted noise above a reference noise level of -90 dBm (defined as 0 dBrnc0). Typical objectives range from 28 dBrnc0 for long-haul routes to 34 dBrnc0 for short-haul routes. When the signal fades the noise increases until the threshold noise level is reached. When the threshold (typically 55 to 58 dBrnc0) is exceeded, the transmission quality is considered unacceptable.

With digitally modulated systems the bit error rate (BER) is the measure of transmission quality. The bit error rate is the number of bit errors per total received averaged over a period of time. If the transmission rate is 10 Mbits per second and 100 bit errors occur over a 100 second period, the BER is 10^{-7}, an average of 1 error in 10^6 bits. The acceptable level of transmission is determined by the type of traffic.

For PCM voice traffic, bit errors manifest themselves as clicks and a threshold of 10^{-6} (1 click approximately every 15 seconds) is usually considered acceptable. At this threshold the speech is intelligible, but beyond it the clicks become annoying and intelligibility falls rapidly.

For data with error correction a higher BER of 10^{-8} is normally acceptable (Communications International, 1989).

11.6 Terminology

A number of different units are used worldwide to define the performance and transmission levels of a radio relay system (for decibel definitions see Chapter 2). Important international definitions and units are:

- *Zero transmission reference point.* This is a point arbitrarily established in a transmission circuit, with all other levels in the circuit being stated with reference to this point. Its relative level is 0 dBr.
- *Standard test tone.* The standard test tone for use at audio circuit points is defined as a power of 1 milliwatt (0 dBm) at a frequency of 1000 Hz applied at the zero transmission reference level point.

11.7 Link planning

Planning a link involves producing a profile of the path and calculating the net loss in the system to arrive at a transmitter output power which will produce the designed signal-to-noise ratio at the receiver. A simple example using the profile of *Figure 11.2* is shown in Section 11.8.

Transmission lines and waveguides are discussed in Chapter 3 and microwave antennas in Chapter 4.

11.8 Example of microwave link plan

Frequency: 2000 MHz

Antenna type, station A: P6F-17C	height agl.	20 m
Antenna type, station B: P6F-17C	height agl.	20 m
Feeder type, station A: LDF5P50A	loss, dB/100 m	6.46
Total length, antenna to equipment		30 m
Feeder type, station B: LDF5P50A	loss, dB/100 m	6.46
Total length, antenna to equipment		30 m

Performance

Path length: 20 km, therefore clear path loss	124.0 dBi
Obstruction loss	0.0 dB
Feeder loss, station A:	1.9 dB
Feeder loss, station B:	1.9 dB
Feeder tail loss, total for link	1.5 dB
Connector loss, total for link	1.5 dB
Total loss	**130.8 dB**
Gain, antenna A	28.6 dB
Gain, antenna B	28.6 dB
Total gain	**57.2 dB**
Nett loss (total loss − total gain)	**73.6 dB**
Receiver threshold for max. signal/noise	−125.0 dBW
Design fade margin	+30.0 dB
Design receiver input level	**−95.0 dB**
(threshold − fade margin)	
Transmitter output power	**−21.4 dBW**
(receiver input − nett loss)	

Reference

Krzyczkowski, M. (1989). *Communications International*, August.

12 Information privacy and encryption

12.1 Encryption principles

Radio communication was never secret, but since the advent of fast frequency scanning receivers the ability to overhear, even on cellular radio telephones, is within easy reach of anyone. Privacy systems are, however, available which will deter the casual listener and gain time against the determined eavesdropper.

Messages, which may be either speech or data, are encrypted to prevent both eavesdropping and the injection of spurious information. The aim is to make the encryption and decryption as easy and inexpensive as possible for authorized users and time-consuming, difficult and costly for the eavesdropper (cryptanalyst). *Figure 12.1* shows a standard cryptographic channel.

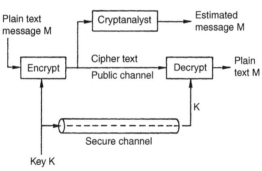

Figure 12.1 Cryptographic channel

A plain text message M (speech, written or digital) is encrypted by mixing with a key K to produce a cipher text. The cipher text may be transmitted over a channel which is accessible to the public and hence to the cryptanalyst. The key is issued via a secure channel to the authorized recipient who uses it to decipher the message. The cryptanalyst without access to the key attempts to derive the maximum information from the cipher text to enable him or her to estimate the content of the message.

One key may be used continuously or for long periods or, to increase the cryptanalyst's confusion, the key may be changed frequently, perhaps even for each character of the message. A sequence of key changes which repeats after a fixed number of characters produces what is known as periodic encryption.

Encryption may be either symmetrical or asymmetrical. Symmetrical encryption uses the same key for both encryption and decryption. Asymmetrical encryption uses a different key for each process, thus providing for different levels of authorization. Encryption keys may be supplied to many persons who are authorized to transmit encrypted messages but decryption keys may be issued to only a few authorized recipients.

12.2 Speech encryption

The encryption of speech offers fewer possibilities than does the encryption of written or digital data messages. The simplest method of encrypting speech is scrambling by inverting the speech frequencies; *Figure 12.2* shows this process.

Figure 12.2 Speech inversion

The speech, contained in the band 300–3400 Hz, is mixed with a key frequency of 3700 Hz producing an erect, upper side band from 4000–7100 Hz, and an inverted, lower side band where the 300 Hz components of speech have become 3400 Hz and the 3400 Hz have been inverted to 300 Hz. The upper side band is rejected by a lowpass filter and the inverted lower side band is transmitted. In the receiver the scrambled speech is mixed again with 3700 Hz to produce an erect side band – the original non-inverted speech message.

Simple inverted speech is easily unscrambled. There is little choice of key frequency and if the eavesdropper uses a slightly different frequency the pitch is changed but the speech is readable. Also, if the centre frequencies only of the inverted band are selected by means of a band-pass filter, inverted speech becomes intelligible.

A sophistication which renders the speech more secure divides the speech band into sections and transmits them separately using a different key frequency for each band (audio frequency hopping). The divisions of the speech band may also be treated as blocks and transposed in time according to a user-programmable pattern to create further confusion in the mind of the eavesdropper.

The most up-to-date methods of speech scrambling convert the speech into digital form by either pulse code modulation (PCM) or some other method. The digits corresponding to the speech may then be either transmitted as frequency modulation, e.g. FFSK, on an analogue radio system or, possibly after further encryption, transmitted directly on a digital system. Digitized speech creates improved security not only by the digitization itself but by offering the higher encryption capabilities of data.

The price to be paid for security with analogue encryption is a degradation of the received signal-to-noise ratio by 9 dB, effectively reducing the range of a transmitter by approximately 40%.

12.3 Data encryption

Digital data may be encrypted by changing the digits, perhaps adding superfluous digits, and transmitting the resultant cipher message either in blocks of a fixed size or as a stream.

Block encryption treats the blocks in the same way as in the encryption of speech, with different keys being used for each block – or each character – and the blocks re-distributed in time.

Stream encryption has no fixed block size and each plain text bit, M_i, is encrypted with the ith element, K_i, of a pseudo-random, sometimes called pseudo-noise (PN), key.

Figure 12.3 shows two methods of generating pseudo-random keys or pseudo-noise. The first, *Figure 12.3(a)*, operates as follows. At each clock pulse the contents of the pre-loaded four-stage shift register are stepped forward from left to right. Immediately after the shift, the output bit is fed back into the 1st and 2nd stages. It introduces a new bit into stage 1, and is added by modulo 2 addition to the new content of stage 2, producing a new set of contents. The initial loading of 1100 emerges as a pseudo-random 15 bit sequence which then repeats. The periodicity of the sequence is given by:

$$\text{Sequence length, characters} = 2^N - 1$$

where $N =$ the number of stages in the shift register.

170

(a)

(b)

Code length $= 2^N - 1$
e.g. $N = 5$, length $= 31$ characters

Figure 12.3 Pseudo-noise (PN) generation

For a four-stage register, therefore, the sequence repeats after 15 bits and is shown in *Figure 12.3(a)*.

A more commonly used method combines the outputs of two or more of the earlier stages in a modulo 2 adder and feeds the result back to the input of the register as in *Figure 12.3(b)*.

To form the cipher text the resultant pseudo-random key is mixed with the original data message in a second modulo 2 adder. If the clock rate for the shift register is the same as the bit rate of the plain text message, the plain text bits are exchanged for those of the modulo 2 sum, but if the shift register runs faster than the plain text bit rate, additional bits are added into the cipher text. This is more common and extends the time taken by a cryptanalyst to estimate the message. The price to be paid for the improved security is either a slower effective bit rate for the message or a higher overall bit rate and hence an increased bandwidth requirement.

Mixing the cipher text with the output from an identical PN generator in the receiver recovers the original text.

Modulo 2 addition

A modulo 2 adder is an exclusive-OR gate which produces a logic 1 output whenever either of the inputs is at logic 1 and the other is at logic 0. When both inputs are identical, the exclusive-OR gate produces a logic 0 output.

The truth table for an exclusive-OR gate is:

Input		Output
A	B	Y
0	0	0
0	1	1
1	0	1
1	1	0

Modulo addition is not limited to two inputs. Any quantity of binary numbers may be added: if there is an odd number of logic 1s in a column, the adder produces a logic 1 output, if an even number, i.e. no remainder in the binary addition, the output is logic 0.

Synchronous encryption

The key is generated independently of the message from a previously loaded register. If a character is lost during transmission of a synchronous text, resynchronization of transmitter and receiver key generators is necessary before transmission can continue.

Self-synchronous encryption

The key for each character is derived from a fixed number of previous characters of the plain text message. For example, the shift register is pre-loaded with the plain text characters so that in a four-stage register the key used for encrypting the 4th character will be the 4th previous message character. If a self-synchronized transmission loses a character, the system automatically re-synchronizes the same number of characters (in this case four) later.

Written messages may be encrypted using one of the classical mechanical methods of rearrangement of the letters before digital encryption.

When sufficient RF channels exist frequency hopping is a further possibility, and the spread spectrum technique, where the signal energy is spread over a very wide band of frequencies, not only offers very high security but also makes detection of the signal difficult. The shift register techniques described above are also used for generation of the frequency hopping sequence and the spreading of the base band frequencies.

12.4 Code division multiple access (CDMA) or spread spectrum

The extension of pseudo-random key or noise generation is code division multiple access or spread spectrum transmission, described in Chapter 8. The spread spectrum technique provides an extremely high level of security by reducing the radiated energy at any one frequency to very little above the ambient noise level by spreading the transmission over a very wide band. The transmitter uses what is in effect an extended digital key to spread the bandwidth and the receiver is equipped with an identical key for de-spreading. The transmission almost disappears into the noise and, without the appropriate key, the existence of a spread spectrum signal is very difficult to detect.

12.5 Classification of security

Unconditionally secure

Those systems where the cryptanalyst has insufficient information to estimate the content of the cipher regardless of the amount of time and computation facilities available. This is only realistic when using what is known as a one-time pad where the key is used once and once only.

Computationally secure

Encryption systems are specified in terms of the amount of time taken by a cryptanalyst to estimate the cipher's contents using the state of the art techniques. Unless an extremely long periodicity is used for a progressive key – months and even years in some instances – requiring many stages in the shift register, it is possible for a cryptanalyst who knows, or can estimate, a small part of the message to calculate all the parameters necessary to decipher the message. However, stream encryption with a pseudo-random key approaches perfect secrecy for a finite number of messages.

References

Chambers, W.G. (1985). *Basics of Communications and Coding*. Oxford University Press, Oxford.

Sklar, B. (1988). *Digital Communications*. Prentice Hall, Englewood Cliffs, NJ.

13 Multiplexing

Multiplexing enables several information (speech or data) channels to be carried simultaneously over one bearer, a wide band, single frequency microwave radio link for example. Both frequency and time division multiplex are common methods. On trunked radio systems where channels are allocated to users on demand the multiplexing is referred to as frequency- or time-division multiple access (FDMA, TDMA).

Frequency division multiplexing (FDM) used to be a common technique for sharing analogue trunk circuits between 12 (or multiples of 12) separate channels. This is an analogue technique that been superseded by the use of digital trunk circuits, where digital multiplexing is employed. FDM is still used on some satellite and microwave links, although most links are now using digital techniques. Some systems such as cellular telephones use separate transmit and receive frequencies, which is known as frequency division duplexing (FDD).

13.1 Frequency division multiplex

Frequency division multiplex divides a broad band of frequencies into slots, each wide enough to accommodate an information channel. This is achieved by amplitude modulating a higher frequency subcarrier with each speech signal to form groups of channels.

Each speech channel contains frequencies between 300 and 3400 Hz plus, in some systems, an out-of-band signalling tone of 3825 Hz and a guard band. Each channel modulates a base band subcarrier spaced at 4 kHz intervals upwards from 64 kHz. This produces an upper and lower side band from each channel (*Figure 13.1*). The carrier and upper side band are removed by filters and the lower (inverted) side band is transmitted. At the receiver, the base band frequencies are again mixed with the same subcarrier frequency to restore the original speech. The subcarrier frequencies are maintained to an accuracy of ± 1 Hz which creates the ± 2 Hz frequency translation error quoted in some telephone line specifications.

Twelve such channels form a ITU-T basic group B occupying the band between 60 and 108 kHz. This basic group B may now be

(a) Production of side bands in frequency division multiplex

(b) Production of basic group B

(c) Production of groups A and B

Figure 13.1 Frequency division multiplex

mixed with 120 kHz to produce a lower side band of 12 to 60 kHz, now basic group A. Filters leave 60 to 108 kHz free for a new basic group B.

The process may be repeated by using five basic groups to modulate still higher-frequency carriers, to produce super- and hypergroups.

For FDM data communications, the bearer circuit bandwidth of 3000 Hz is divided into 12 channels each of 240 Hz bandwidth. Data is transmitted at 110 bits per second allowing a send-and-receive channel in each block of 240 Hz.

13.2 Time division multiplex (TDM)

A time division multiplex system conveys digital data, and speech must first be converted to data. TDM allocates short-duration time slots within a wider time frame to each information channel. For example, a continuous stream of data sent over a link at a rate of 2400 bit/s could convey the information contained in four 600 bit/s channels in short sequential bursts.

If the duration of one input bit is 1/600 s or 1.666 ms, a seven-bit character occupies 11.66 ms and 85 such characters can be sent per second. If the transmitted rate can be speeded up each bit sent at 2400 bit/s has a duration of 416 μs and 343 seven-bit characters can be sent per second. Such a system is shown in *Figure 13.2*. The data is stored in the buffers at the transmitter and the clock pulses are applied to each store/gate sequentially allowing one character from each data channel to be transmitted at a rate of 2400 bit/s. Perfect synchronization must be maintained between all channels and the transmitter and receiver to avoid data errors.

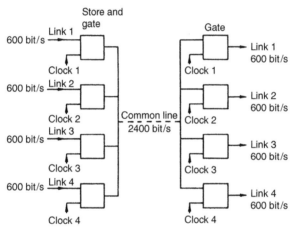

Figure 13.2 Time division multiplex

The principle is illustrated in *Figure 13.3* where a 1 s time frame at 2400 bit/s contains 343 time slots, each of 2.92 ms and containing a character from a specific information channel. Every second, therefore:

channel A would occupy slots 1, 5, 9, ..., 337

channel B would occupy slots 2, 6, 10, ..., 338

channel C would occupy slots 3, 7, 11, ..., 339

channel D would occupy slots 4, 8, 12, ..., 340

This leaves three blank slots; in practice slots are also allocated for preamble, address and synchronization purposes.

Information channels

Figure 13.3 TDM time frame

13.2.1 European E1 multiplexing

The digital base band signal adopted in Europe operates at 64 kbit/s and the multiplexed signal at 2048 kbit/s. An eight-bit word or sample of a PCM voice channel (see Chapter 14) occupies $3.9\,\mu s$, and the interval between successive samples of a channel is $125\,\mu s$, the time frame duration. Therefore, the number of channels (time slots) that can be accommodated in one frame is $125/3.9 = 32$. Thirty of the slots are used for information channels and two for control purposes (*Figure 13.4*).

Figure 13.4 TS0, TS16 are used for signalling; TS1–TS15 and TS17–TS31 are used for voice channels

In the European E1 system, timeslots are numbered 0 to 32. Timeslot zero (TS0) is used for synchronization and TS16 is used for signalling. The signalling channel is therefore operating at 64 kbit/s. Digitized voice is carried in the remaining 30 timeslots. Voice channels 1 to 15 are carried over TS1–15. Voice channels 16–30 are carried over TS17–31.

13.2.2 American T1 (D1) multiplexing

The American T1 system is a 1.544 Mbit/s transmission system. The data format should be referred to as DS-1, but T1 is sometimes used instead. Timeslots (TS), or channels, in this system are numbered 1 to 24; there are 23 available to carry traffic and TS24 is used for synchronization. Each timeslot carries 8 bits, sampled at 8 kHz. An additional bit is sent at the end of each frame (known as the frame-bit or F-bit) so there are 24, 8-bit timeslots, plus 1 bit = 193 bits per frame.

The synchronization word sent in TS24 is coded as 1 0 1 1 1 Y R 0. The Y and R are dependent upon the data multiplexer and are for the manufacturer's use.

Signalling information is transmitted using 'stolen' bits from each channel, rather than using a separate timeslot as in the European E1 system. Bits are not stolen from timeslots in every frame, because this would degrade the quality of a voice channel. Instead, the least significant bit of every sixth frame is used for signalling. Thus the effective word length of the encoded voice is 7.833 bits, rather than 8 bits of the E1 system.

There are two virtual signalling channels (A and B) created by the use of stolen bits. Signalling Channel A is derived from bits stolen from timeslots during the sixth frame. Signalling Channel B is derived from bits stolen from timeslots during the twelfth frame. This is repeated for every superframe. Each channel has a data rate of 24 bits per 12 frames (1.5 ms), or 16 kbit/s, giving 32 kbit/s signalling in total. Although this data rate is half that of the E1 system, there are only 23 voice channels instead of the 30 used in the E1 system.

The F-bit is used to identify the frame, which is important when extracting the signalling information. Because only one in six frames are used for signalling, it is vital to know which ones carry the information and which ones carry voice traffic. Also there are two signalling channels and these must be separated at the receiver end of the system. A six bit data pattern of 1 0 0 0 1 1 is transmitted, followed by the inverse: 0 1 1 1 0 0. The complete 12-frame sequence is known as a superframe.

13.3 Code division multiple access (CDMA)

Spread spectrum transmission (described in Chapter 8) is a form of multiplexing. In addition to high security it permits multiple occupation of the wide – typically 1.25 MHz – frequency band. A number of users possessing keys of low correlation can occupy the same band at the same time. The system operates well in poor signal-to-noise or high interference environments.

Digital signals can be multiplexed by coding each source individually. Instead of transmitting logical 1's and 0's, a data pattern is assigned to each bit. The pattern for a logical 1 could be a ten-bit word (e.g. 1011010011); a separate pattern would represent a logical 0. Correlation (i.e. pattern matching) between a known signal and a signal received with a certain code gives an indication of whether the bit received is a 1 or a 0. This technique allows the occasional error in the data path without affecting the received bit. More often, it allows several users to transmit at the same frequency. Each user is allocated a different data pattern so that only the correct person will be able to decode the signal; all other signals will be ignored.

Reference

Winder, S.W. (2001). *Newnes Telecommunications Pocket Book*. Butterworth-Heinemann, Oxford.

14 Speech digitization and synthesis

14.1 Pulse amplitude modulation

The digitization of analogue waveforms by pulse code modulation is accomplished in two stages. First the waveform is sampled to produce pulse amplitude modulation (PAM). Short-duration samples are taken at regular intervals which are long compared with the sampling time but short in relation to the highest signal frequency. The result is a train of pulses whose amplitude envelope is the same as the analogue waveform. The envelope (*Figure 14.1*) will contain:

- clock frequency f_c, the sampling rate
- all the signal frequencies contained in the waveform from the lowest, f_1, to the highest, f_2
- upper and lower side bands, $(f_c - f_2)$ to $(f_c - f_1)$ and $(f_c + f_1)$ to $(f_c + f_2)$
- harmonics of f_c and the upper and lower side band frequencies
- a DC component equal to the mean value of the PAM waveform.

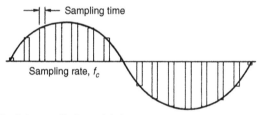

Figure 14.1 Pulse amplitude modulation

The envelope contains the original signal frequencies and can be demodulated by a low-pass filter which will pass f_2 but not the clock frequency. The clock frequency must therefore be higher than $2f_2$. For line communications in the UK a clock frequency of 8 kHz is used with a maximum modulating frequency of 3400 Hz.

14.2 Pulse code modulation

To overcome the susceptibility of PAM signals to corruption of the amplitude waveform by noise and interference, the waveform is processed further to produce pulse code modulation (PCM) before transmission.

180

In this process the magnitude of the PAM samples with respect to a fixed reference is quantized and converted to a digital (data) signal. Quantizing rounds off the instantaneous sample pulse amplitude to the nearest one of a number of adjacent voltage levels. *Figure 14.2* illustrates the process for an eight-level system. In the figure the amplitude at t_0 is 2, between 4 and 5 at t_1, between 5 and 6 at t_2, etc. After quantization the values would be 2 at t_0, 5 at t_1, 5 at t_2 and so on. The difference between the amplitude levels and the rounded-off values is the quantization noise or distortion.

Figure 14.2 Eight-level pulse code modulation

The binary pulse train – leaving a one-bit synchronizing space between each number – for this example would be as in *Figure 14.3*.

s = Synchronization space

Figure 14.3 Binary pulse train (from *Figure 14.2*)

The number of quantizing levels is 2^n and the highest denary number represented is $(2^n - 1)$ where n is the number of bits used to represent each sample. If each train of pulses is accompanied by one synchronizing bit, the number of bits per sample is $(n + 1)$. If the sampling rate is f_c, the transmitted bit rate is $(n + 1)f_c$. For example, an eight-bit word – including the synch. bit – is commonly used to represent a sample, so, with a clock frequency of 8 kHz:

$$\text{number of quantizing levels} = 2^7 = 128$$

$$\text{sampling rate, } f_c = 8\,\text{kHz}$$

$$\text{therefore transmitted bit rate} = (n + 1)f_c$$

$$= 64\,\text{kb/s}$$

The maximum frequency of the pulses will be when transmitting alternate 1s and 0s, and the occupied bandwidth, $\frac{1}{2} \times$ bit rate, 32 kb/s.

Quantization as described above is linear, i.e. the spacing of the quantization levels is the same over the range of pulse amplitudes. This produces a poor signal-to-noise ratio for low-level signals which is improved by the use of more closely-spaced levels at small signal amplitudes than at large amplitudes. Where non-linear quantization is used the most significant bit of the sample character identifies the polarity of the signal.

If linear quantization is used then about 12 bits per sample are needed to give a good voice quality, which requires a bit rate of about 96 kbits/s. For coding speech using non-linear quantization 8 bits per sample is sufficient, which requires a bit rate of 64 kbits/s. Two slightly different non-linear PCM codecs were standardized in the 1960s. In America u-law coding is the standard, but in Europe A-law coding is used.

14.3 ADPCM codecs

Adaptive differential pulse code modulation (ADPCM) codecs are waveform codecs which quantize the difference between the speech signal and a prediction of the speech signal. If the prediction is accurate, then the difference between the real and predicted speech samples will be small. The signal will be accurately quantized with fewer bits than would be needed using standard PCM. At the decoder the quantized difference signal is added to the predicted signal to give the reconstructed speech signal.

The performance of the codec is aided by using adaptive prediction and quantization. The predictor and difference quantizer adapt themselves to the changing characteristics of the speech being coded.

The CCITT (now ITU-T) standardized a 32 kbits/s ADPCM, known as G721, in the 1980s. The quality of the reconstructed speech was almost as good as that from standard 64 kbits/s PCM codecs. Since then G726 and G727 codecs operating at 40, 32, 24 and 16 kbits/s were standardized.

14.4 The G728 low delay CELP codec

At bit rates of around 16 kbits/s and lower the quality of waveform codecs falls rapidly. Thus, at these rates, CELP codecs and their derivatives, tend to be used. CELP stands for Code Excited Linear Prediction. A code representing a voice pattern is transmitted, rather than coding short samples of voice signal. The code is calculated by looking at the long-term changes in the voice and then predicting how the voice pattern will change. Since the pattern takes some time to establish, CELP codecs produce a delay.

A backward adaptive CELP codec was standardized in 1992 as G728. This codec uses backward adaptation in its calculations. This means that, rather than buffer 20 ms or so of the input speech for its calculations, the pattern of previous speech is used. This means that the G728 codec has to analyse fewer samples than traditional CELP codecs. The G728 codec uses only 5 samples giving it a total delay of less than 2 ms.

14.5 The GSM codec

The 'Global System for Mobile communications' (GSM) is a digital mobile radio system which is extensively used throughout Europe, and also in many other parts of the world. The GSM codec provides good quality speech, although not as good as the slightly higher rate G728 codec.

The GSM full rate speech codec operates at 13 kbits/s and uses a regular pulse excited (RPE) codec. This codec has short-term and long-term predictors to estimate how the voice pattern will change over time.

The input speech is split up into 20 ms long sections, called frames, and for each frame a set of eight filter coefficients are calculated for the short-term predictor. Each frame is then divided into four 5 ms sub-frames, and for each sub-frame the encoder calculates the delay and gain needed for the codec's long term predictor. Finally, after both short and long term prediction has been applied, the residual signal is quantized for each sub-frame. The filter coefficients for the short term predictor, the delay and gain of the long-term predictor and the coded residual signal are all transmitted. The receiver uses all these values to reconstruct the signal.

References

Kennedy, G. (1977). *Electronic Communications Systems*, McGraw-Hill Koga-shuka, Tokyo.

Winder, S.W. (2001). *Newnes Telecommunications Pocket Book*, Butterworth-Heinemann, Oxford.

15 VHF and UHF mobile communication

Mobile communication operating throughout the VHF and UHF bands is expanding rapidly and, although the fastest growth has been in the radio-telephone field, interest in private mobile radio (PMR) is undiminished. Most of the procedures described for PMR are applicable to many other branches of radio communication, so traditional PMR is considered here. Other systems are discussed in later chapters.

A private mobile radio system, comprising a base station and mobiles, is one that is effectively owned by the user and, under the conditions of the licence, may only be operated by the user's own staff for his or her own business. Airtime cannot be leased to other persons.

15.1 Operating procedures

Frequencies are normally allocated in pairs, one for the up-link to the base station and one for the down-link to the mobile (not to be confused with a radio link used for control purposes). Such a pair of frequencies, spaced sufficiently apart to permit simultaneous transmission and reception by a station, comprises the radio channel. Occasionally, for special purposes and for small, low-power, and possibly temporary, operations a single frequency only may be allocated. The mountain rescue teams are an example.

The methods of operating are as follows.

Single-frequency simplex

This method uses a common frequency for transmission and reception by all stations operating on the system (*Figure 15.1*). Transmission

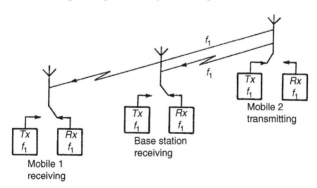

Figure 15.1 Single-frequency simplex

and reception cannot take place simultaneously at a station, and a receiver is switched off whenever the transmit switch is operated. This prevents blocking of the receiver by the transmitter and acoustic feedback occurring. The method allows all stations within range to hear both sides of a conversation and to relay messages to more distant stations; an obvious advantage for mountain rescue.

At the end of a transmission an operator must say 'over' and switch off the transmitter to hear the reply. A conversation must end with the word 'out' so that other stations are aware that the system is unoccupied.

Two-frequency simplex

Separate frequencies are used for transmission and reception but whilst a station is transmitting its receiver is still switched off (*Figure 15.2*). Mobiles hear only the base station and, therefore, the relaying of messages is not possible. A further disadvantage is that because mobiles hear only the base station, they may be unaware of the system occupancy and transmit, interfering with an existing conversation. The advantage of two-frequency simplex is the avoidance of receiver blocking or de-sensitization, not only from the associated transmitter but also, at base stations where several channels within the same band are located, from nearby transmitters.

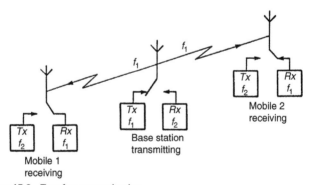

Figure 15.2 Two-frequency simplex

Duplex and semi-duplex

Separate frequencies are used for transmission and reception and, in full duplex, all stations can transmit and receive simultaneously as in a two-way telephone conversation (*Figure 15.3*). While a station is transmitting, its receiver audio output is switched from the loudspeaker

Figure 15.3 Duplex

to an earpiece to prevent acoustic feedback. A mobile cannot receive other mobiles directly but full duplex enables all stations to break in on a conversation in an emergency or to query part of a message; it also facilitates the use of talk-through where mobiles can speak to each other via the base station. To maintain awareness of system occupancy the base station may transmit a series of pips as an engaged signal during pauses in the despatcher's speech.

Many systems operate semi-duplex where only the base station operates a duplex procedure and the mobiles use a simplex procedure. This avoids the higher cost of duplex mobiles and offers most of the facilities of duplex, except that a despatcher cannot break in on a transmitting mobile.

Open channel/'All informed'

All mobiles hear all the calls from control, i.e. no selective calling is in operation.

Selective calling

Mobile receivers remain quiescent until specifically addressed; the opposite of open channel working. Individual mobiles, groups and a whole fleet may be addressed.

Auto-acknowledgement

When selectively called a mobile automatically transponds, sending the code for its address and, possibly, its status information. Mobiles can only acknowledge when individually addressed; auto-acknowledgement on group and fleet calls is prevented to avoid mobiles transponding simultaneously.

Status updating

The transmission, automatically or manually, of the data denoting the mobile's current status.

Call stacking

The storage of calls from mobiles and their presentation in call order to the despatcher. Arrangements are usually made to raise urgent calls to the top of the stack with an enhanced display.

15.2 Control of base stations

Where adequate radio signals over the desired service area of the system can be provided by one base station sited at the control point, it can be easily controlled, either directly from the front panel or over a multi-way cable.

In many instances, to obtain adequate coverage, the base station must be sited on high ground remote from the control point. Then, either a land line or a radio link, both of which will today probably be digital and the link microwave, must be used for control. A land line will most likely be rented but radio links are favoured by those users who insist on the complete system being under their direct control.

15.3 Common base station (CBS) operation

An economic method of providing mobile communication for users with a small quantity of mobiles and light traffic is a common base station or community repeater system. Although still referred to as private mobile radio, the base station is shared by several users who pay a fixed subscription for the service irrespective of airtime used. The station is controlled by radio using tone-controlled talk-through or what is sometimes called reverse frequency trigger (not normally permitted on a true PMR system). *Figure 15.4* shows the layout of a single-station CBS system. Each participant's office contains a fixed transmitter/receiver operating on the same frequencies as a mobile and equipped with CTCSS. Different CTCSS tones are assigned to each user to ensure privacy. All office stations must use a directional antenna so that they access only the base station to which they subscribe.

When a user transmits, from either the office station or a mobile, the base station, on receipt of a signal containing a valid CTCSS tone,

Base transmitter
and receiver

User B's fleet

User A's fleet

Directional
antenna

Despatcher's
VHF transmitter/
receiver

User A

User B

Figure 15.4 Common base station system

enters a talk-through mode. The caller's CTCSS tone is retransmitted and, in turn, opens the mute on all that user's mobiles allowing communication via talk-through. Mobiles are equipped with 'busy' lamps to indicate whenever the system is engaged, and transmission time-out timers to prevent excessively long calls excluding other users.

15.4 Wide area coverage

Where the desired service area for one radio channel is larger than can be covered by a single base station the way in which the base stations are to be controlled requires special consideration. Crucial aspects are the presentation of the best received signal from a mobile to the control operator or despatcher and the selection of the transmitter most likely to provide the best signal at the location of the mobile.

Receiver voting

Well-proven receiver voting circuits which present the best received signal to the despatcher have been used for many years. These circuits sample the signals received from a mobile at each base station and, by means of coded information – which may be either digital or in the form of continuous tones – enable equipment at the control centre to automatically select the best. The selection may be made by comparing

188

either the signal-to-noise ratio or the signal strength of the received signal. If the information is to be used solely to select the best signal in terms of readability, the signal-to-noise ratio is probably the better characteristic to use, but if the information is also to be used to select a base transmitter, the signal strength could be considered more satisfactory. Some systems utilize both types of information.

A typical 3-station voting system is shown in *Figure 15.5* where signal sampling and vote encoding occur at the receiving sites and the coded information is passed over the base station control system to the control centre. This method is necessary when the selection is made on a signal strength basis, but where the signal-to-noise ratio is used for the selection the sampling and encoding can be done at the control centre taking into account the noise occurring in the control network. Receiver voting systems operate very quickly, and changes

Figure 15.5 Radio scheme with one transmit/receive plus two receiver-only stations with voting

of the selected receiving site may occur several times during a message without the despatcher's awareness.

The broadcast transmit and receive paths are not always reciprocal, for instance when low power hand-portables are integrated with higher power vehicular mobiles. In these circumstances the use of additional receiver-only fill-in stations is an economic and satisfactory proposition.

Base transmitter control poses a problem much more difficult to resolve. The selection of a base transmitter to communicate with a mobile whose precise whereabouts may be unknown, the broadcasting of messages to all, or groups of, mobiles whose locations may be widespread, and the provision of talk-through between mobiles are all facilities required on major systems, and difficult to provide satisfactorily. Apart from trunking, which can be economic only on very large networks, there are three traditional methods of operating the base transmitters on wide area schemes: manual selection, automatic selection and simultaneous transmission from more than one transmitter.

Manual transmitter selection

On many systems the despatchers select the transmitters manually. It is the simplest and least expensive method but has serious disadvantages:

1. Making the selection may entail trying a number of transmitters before sending the message, increasing the operator's work load and wasting air time.
2. Mobiles outside the service area of the selected transmitter may call, interrupting an existing conversation, either because they are unaware that the system is engaged or have received a poor signal that they believe may have been intended for them. Transmitting bursts of 'engaged' pips sequentially over the unselected transmitters during pauses in the despatcher's speech alleviates the first situation; the use of selective calling overcomes the second.
3. Broadcast messages must be transmitted on each transmitter in turn and talk-through between mobiles which are not in the service area of the same transmitter is not practicable.

Automatic transmitter selection

Selecting the transmitters automatically, or semi-automatically, is an improvement over manual selection. On a system where the mobiles are not equipped with selective calling and automatic acknowledgement of a call, automatic selection of transmitters can only occur on receipt of a call from a mobile. The transmitter through which to reply is then selected by the receiver voting system at the same time as it selects the signal to present to the despatcher. The selection is made at the start of the call and, because of the switching times involved, the transmitter selected is retained for the duration of the call. On these open channel systems, a calling despatcher must initially manually select a transmitter.

On systems where selective calling and auto-acknowledgement of a call are provided, the system can be made fully automatic by transmitting the data corresponding to a mobile's call sign from each transmitter in turn until a satisfactory acknowledgement is received. The successful transmitter is then retained for the duration of the conversation and, at its conclusion, is usually the one used to commence another call.

Such a system overcomes the disadvantage of the need to manually select the transmitter but the problems with mobiles outside the service area of the selected transmitter, and of broadcasting and talk-through, remain.

Simultaneous transmission

Operationally, simultaneous transmission from all sites is ideal and, under various names such as Spaced Carrier, Simulcast and Quasi-synch., has been around since the mid-1940s. Its operational value is proven but systems require special care in their planning, adjustment and subsequent maintenance.

An early form of simultaneous transmission was the amplitude modulated spaced carrier system. Used very successfully at VHF on systems using 25 kHz channel separation, the transmitter carrier frequencies were separated by 7 kHz – above the mobile receivers' audio pass-band. With the reduction of channel spacings to 12.5 kHz, spaced carrier operation on this basis was no longer possible and alternatives are to either synchronize, or very nearly synchronize, the carrier frequencies. There are, however, undesirable effects of synchronous and quasi-synchronous transmission but, with care, these can be reduced to an acceptable level. They are:

1. The beat note between transmitters being audible in the mobile receiver.
2. Variations in signal level due to interference patterns between signals from more than one transmitter.
3. Distortion due to audio phase differences and differing modulation levels when signals of comparable strength are received from more than one transmitter.

The beat note is easily dealt with. It is rendered unobjectionable either by placing it outside the mobile receiver audio pass-band – which is the usual method – or by synchronizing the transmitter carrier frequencies so that no beat note is produced. Synchronization, however, raises other problems, which are particularly severe at VHF but less so at UHF.

The spaced carrier system placed the beat note above the receiver pass-band, but modern systems place it below. This means that the beat note between the lowest and highest carrier frequencies must be below about 150 Hz if it is to be unobtrusive. Tests have shown the optimum carrier separation to be from 0.5 Hz to 4 Hz between any two transmitters using amplitude modulation, and from 5 Hz to 40 Hz between adjacent frequency transmitters using angle modulation.

Two transmitters on the same or closely-spaced frequencies produce deep nulls in the received signal in the areas where they provide almost equal strength signals. This is a natural phenomenon and the effect can only be reduced by the correct siting of stations and antenna configurations. Because of the longer wavelength the effect is more detrimental at VHF than UHF. Where the frequencies are quasi-synchronous the interference pattern is continually moving, but with synchronized carriers the pattern is virtually stationary and at low band VHF wavelengths it is possible to stop a vehicle in a place of semi-permanent zero signal. While moving slowly in an area of equal signal strengths from two transmitters, the cancellations become very objectionable. At 450 MHz the distance between the nulls is so short that it is almost impossible to remain in one, and while moving they are unnoticed. Strong signals in the overlap areas minimize the time that the signals fall below the receiver noise threshold at each cancellation. They are the key to reducing the annoyance from the cancellations. Amplitude modulated systems have an advantage in that the receivers produce less severe bursts of noise during the signal nulls.

The audio distortion, provided the equipment does not introduce significant additional harmonic distortion, is attributable to audio phase differences in the signals received from more than one transmitter. For the distortion to be severe, the received signals must differ by less than 6 dB on an angle modulated system; capture effect in the receiver removes the audible effects at greater differences.

The phase differences arise from differing audio characteristics in all the circuits in the path including land lines, radio links and control equipment. Common to all systems are the phase differences due to the different path lengths from the control centre to a receiving mobile, but where land lines or multiplexed circuits are used for the control

192

of the base stations, variable bulk and group delays and frequency translation errors can present serious difficulties. Tests have shown AM to be slightly more tolerant than angle modulation in respect of this type of distortion, phase delays of 100 μs being acceptable with AM compared with 70 μs with FM.

Differing path lengths between the control centre to the transmitting sites can be equalized by installing audio delay circuits at the transmitters. The path lengths from each transmitting site to all the places where equal signals occur must then be less than about 21 km (equivalent to 70 μs) for acceptable quality. Modern techniques enable delays to be dynamically equalized to compensate for variations in the path.

Figure 15.6 shows the layout of a multi-station scheme including the audio delays. However, signals do not confine themselves to neat

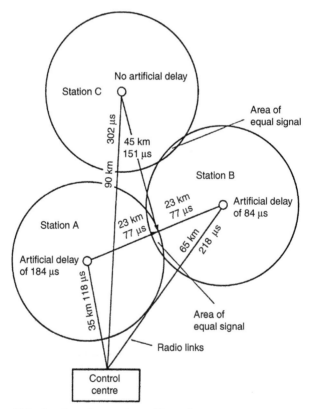

Figure 15.6 Quasi-synchronous transmitter system

circles and the worst situation is where the signals from two transmitters arrive at a mobile more or less equal in strength and phase. In this situation, apart from the ripple caused by the carrier frequency offset of the transmitters, a satisfactory signal would still be received but, occasionally, the presence of a weaker signal from a third, distant station (C in *Figure 15.6*) intrudes during the cancellation periods and, because of its long path length, introduces severe distortion. The area where this situation occurs is often small and the areas of overlap can usually be moved slightly by adjustment of either the transmitter power or antenna directivity.

16 Signalling

Many radio receivers are required to respond to instructions sent to them over the radio channel to which they are tuned. Several methods of signalling these instructions are in use, from the continuous tones used for controlling the receiver mute operation to the rapid data to initiate channel changes and transmitter power control on cellular radio-telephones.

16.1 Sub-audio signalling

The slowest signalling system uses continuous sub-audio frequency tones. Known as continuous tone controlled signalling (previously squelch) system (CTCSS) its performance, in the UK, is specified in Radiocommunications Agency specification MPT 1306. The most common use for the system is to control receiver mute opening. Permitting a mute to open only on receipt of an authorized signal, its use enables privacy between users to be maintained on shared systems, common base station systems for example, and reduces the annoyance factor from interference in the absence of a signal. Thirty-two tones are permitted and assigned by the Radiocommunications Agency (*Table 16.1*).

Table 16.1 CTCSS tones and modulation levels

67.0	110.9	146.2	192.8
71.9	114.8	151.4	203.5
77.0	118.8	156.7	210.7
82.5	123.0	162.2	218.1
88.5	127.3	167.9	225.7
94.8	131.8	173.8	233.6
103.5	136.5	179.9	241.8
107.2	141.3	186.2	250.3

The tones are transmitted at very low modulation levels (*Table 16.2*).

Table 16.2 Modulation levels

System channel spacing	Amplitude modulation Modulation depth (%)	Angle modulation Peak deviation ± Hz
25 kHz	10 to 20	400 to 800
12.5	10 to 20	200 to 400

16.2 In-band tone and digital signalling

16.2.1 Dual tone multi-frequency (DTMF)

Developed for telephone work, dual tone multi-frequency signalling has crept into some radio systems but is slow to operate compared with other in-band systems. Two tones are transmitted simultaneously to represent each digit. In the UK they conform to the MF4 code (*Table 16.3*).

Table 16.3 MF4 code of DTMF tones

Digit	Frequencies (Hz)	
1	697	1209
2	697	1336
3	697	1477
4	770	1209
5	770	1336
6	770	1477
7	852	1209
8	852	1336
9	852	1477
0	941	1336
*	941	1209
#	941	1477

Some push button telephones signal to the local exchange by generating a combination of two frequencies. Each row and each column of the push button keypad is connected to an oscillator of set frequency. When any push button is pressed, tones corresponding to its row and column frequencies are therefore generated. The row and column oscillator frequencies are as shown below:

16.2.2 5-tone signalling

So-called 5-tone signalling uses a series of tones – usually five but other quantities are used in some systems. The system is popular for selective calling and addressing purposes (*Table 16.4*). There are a range of tones established by various organizations but, although the tones in each series may be assigned to each digit, equipment manufacturers sometimes amend the code and the usage. The relevant Radiocommunications Agency document is MPT 1316, *Code of Practice, Selective Signalling for use in the Private Mobile Radio Services*.

Table 16.4 5-tone codes

Standard	ZVEI	DZVEI	EIA	CCIR/ EEA	CCITT	EURO
Tone duration (ms)	70	70	33	100/ 40	100	100
Pause duration (ms)	0	0	0	0	0	0
Code/digit	Tone frequency (Hz)					
0	2400	970	600	1981	400	980
1	1060	1060	741	1124	697	903
2	1160	1160	882	1197	770	833
3	1270	1270	1023	1275	852	767
4	1400	1400	1164	1358	941	707
5	1530	1530	1305	1446	1209	652
6	1670	1670	1446	1540	1335	601
7	1830	1830	1587	1640	1477	554
8	2000	2000	1728	1747	1633	511
9	2200	2200	1869	1860	1800	471
Repeat	2600	2400	459	2110	2300	1063
Alarm	2800	2600		2400		
Free-tone						1153
Group tone	970	825	2151	1055		

ZVEI Zuverein der Electronisches Industrie. Designed to operate on 20 kHz channel spaced systems. On 12.5 kHz channel spaced systems transmission of 2800 Hz creates difficulty and depressed ZVEI (DZVEI) was adopted and recommended in MPT 1316.

EIA Electrical Industries Association.

CCIR Committee Consultatif International Radio Communication. The longer tone duration offers robustness against corruption but is slow to operate. Originally designed for marine use, each digit was transmitted twice to ensure reliability.

EEA Electronic Engineering Association. Recommended in MPT 1316.

CCITT International Telegraph and Telephone Consultative Committee.

16.3 Digital signalling

FFSK

Many analogue modulated systems use digital signalling in the form of 1200 bit/s fast frequency shift keying which is faster than 5-tone but less robust.

DCS

Digitally coded squelch (DCS) is a digital, but slower, alternative to CTCSS. It is not in current use in the UK. DCS continuously repeats the constant binary bit pattern of a 23 bit word. Golay error coding reduces the number of data bits to 12 of which 3 initiate the Golay sequence leaving 9 information bits, i.e. a 9 bit word. With a 9 bit word, $2^9 = 512$ codes are possible but, because of problems with the frequencies produced, the codes with 7 or more 1s or 0s in succession are discarded leaving 104 'clean codes'. The transmitted rate is 134 bit/s and the highest fundamental frequency is therefore 67 Hz. The lowest fundamental frequency is produced when 6 zeros follow 6 ones = 1/6th of the fundamental frequency = 11.7 Hz. The important harmonics are:

$$third = 201\,Hz$$

$$fifth = 335\,Hz$$

$$seventh = 469\,Hz$$

The seventh harmonic is well within the audio pass band of most receivers but at 3 octaves from the fundamental is relatively weak.

16.4 Standard PSTN tones

Tone	UK Frequency	UK Sequence	USA Frequency	USA Sequence
Number unobtainable	400 Hz	Continuous	500 + 600 Hz	Interrupted every second
Number busy	400 Hz	0.75 s on, 0.75 s off	480 + 620 Hz	Interrupted once per second
System busy	400 Hz	0.4 s on, 0.35 s off, 0.225 s on, 0.525 s off	480 + 620 Hz	Interrupted twice per second
Dial tone	50 Hz, 33.3 Hz		350 + 440 Hz	
Ringing tone	400 + 450 Hz, or modulated by 17, 25, or 50 Hz	0.4 s on, 0.2 s off, 0.4 s on, 2 s off	440 + 480 Hz	Interrupted once per second
Pay tone	400 Hz	0.125 s on, 0.125 s off		
Time tones	900 Hz	0.15 s on, 0.85 s off. Three times every three minutes		
Recorder warning			1400 Hz	One burst, every 15 s

References

Belcher, R. *et al.* (1989). *Newnes Mobile Radio Servicing Handbook.* Butterworth-Heinemann, Oxford.
Winder, S.W. (2001). *Newnes Telecommunications Pocket Book.* Butterworth-Heinemann, Oxford.

17 Channel occupancy, availability and trunking

17.1 Channel occupancy and availability

The usefulness of any communications network depends on successful calls. Assuming that no calls fail for technical reasons, the success rate depends on the amount of traffic on the channel and how well it is managed. Channel occupancy is a measure of the amount of traffic carried and is affected by:

1. the number of subscribers (mobiles on a radio network)
2. the number of calls per subscriber in a given time
3. the average duration of each message.

The occupancy of any system varies hourly, daily and possibly seasonally. To be successful the network must be able to cope with the traffic during the periods of highest occupancy and, when considering occupancy it is usual to examine the conditions over one hour, the 'busy hour'. The occupancy of a channel can be expressed as a percentage:

$$\text{Channel occupancy} = \frac{nct}{36}\%$$

where
n = number of subscribers (on a mobile radio system, mobiles on watch)
c = call rate (number of calls per subscriber in 1 hour)
t = average message duration in seconds

Occupancy may also be expressed in erlangs where continuous traffic, i.e. 100% occupancy over a specified period, is 1 erlang. Less than 100% occupancy is expressed as a decimal fraction, e.g. 30% occupancy is 0.3 erlang. Demand may also be expressed in traffic units, A:

$$A = \frac{Ct}{T} \text{ erlangs}$$

where C is the number of calls in time T and t is the mean call duration.

The amount of congestion occurring on a system can be defined by the grade of service it offers. Grade of service is expressed as a percentage in terms of the probability of meeting either congestion or waiting more than a specified delay before a call can be established. Erlang's formula is:

$$\text{Grade of service} = \frac{A^N/N!}{1 + A + A^2/2! + A^3/3! + \cdots + A/N!}$$

where A is the number of traffic units (erlangs) and N is the number of outlets (trunked channels).

The probability that delayed calls will have to wait more than a specified time ($W > t$) is:

$$P(W > t) = \exp\left(-\frac{(C - A)t}{H}\right)$$

where
$A = $ traffic units (erlangs)
$C = $ number of trunked channels
$H = $ mean call duration in busy hour

The above assumes that:

1. the number of users is infinite
2. the intervals between calls are random
3. call durations are random
4. call set-up times are negligible
5. delayed calls are queued so that they are dealt with on a first come, first served basis.

The number of potential radio users is continuously growing and any technique that makes more effective use of the finite radio spectrum is welcome. Whenever technology has permitted, higher frequencies and reduced channel spacings have been employed in the attempt to meet the demand. A method of making more effective use of the available spectrum is trunking.

17.2 Trunking

The generally accepted term for the automatic allocation of communication channels to subscribers on demand is trunking, and a 'trunked radio system' is one in which many users share the use of a common pool of radio channels. Channels from the pool are allocated to users on demand and as they become unoccupied; no channels are allocated to specific users or groups of users.

When making a telephone call, provided the speech quality is good it is immaterial which route the call takes. The same applies to radio; providing the message reaches the right person quickly, the radio frequency which carries it does not matter.

Most trunked radio systems use a calling channel on which calls are requested. When a call is established and accepted a working channel from the pool is allocated for the duration of the call and at its termination the channel is returned to the pool for re-allocation to another user. The important principle is that any user has access to any free channel. This has the advantage that more mobiles per channel can be accommodated, or a better grade of service, i.e. less waiting time for access, can be provided. Basic trunked systems are unsuitable where everyone using the system needs to be a party to every call but the latest trunked mobile radio system arrange, on receipt of a call, to assign a working channel to a particular group of mobiles where they operate within the service area of one base station. They then operate on an 'all informed' basis.

Trunking provides the most benefit when the number of channels required is greater than the number assigned to the system. The effective gain in the number of mobiles that can be accommodated on a trunked system is illustrated by the following example:

A single radio channel with an efficient signalling system offering a 30% grade of service can accommodate 90 mobiles with a mean waiting time of 20 seconds if each mobile makes 1 call. Twenty non-trunked channels cater for 1800 mobiles. A 20 channel trunked system under the same conditions can accommodate 3430 mobiles with a mean waiting time of 16.4 seconds, a trunking gain of over 170 mobiles per channel.

The mobile transmitter/receivers used on trunked systems are more complex than those used on systems where channels are permanently allocated because the mobiles must either scan the channels searching for an available one or one containing a call directed to them, or listen continuously on a channel designated for calling. In the latter case when a call is initiated, a complex 'handshake' procedure occurs and the mobile is automatically switched to the channel allocated for that conversation. The band III public access mobile radio (PAMR) networks use a calling channel and many common base station (CBS) systems use channel scanning methods with no calling channel.

Working channels may be assigned for the duration of conversation, which is message trunking, or for the duration of a transmission, which is transmission trunking. With either type of trunking the channel is not returned to the pool immediately the signal from the mobile is lost; some waiting time is essential to cater for short-term signal dropouts but with message trunking the channel is retained for a longer period of time to allow the complete conversation to take place. On a very busy system this is wasteful and, on a system with very rapid signalling, transmission trunking may be used. Then the use of the working channel is lost very soon after the cessation of transmission and a new working channel assigned on its resumption. Some networks automatically migrate from message trunking to transmission as traffic increases.

17.3 In-band interrupted scan (IBIS) trunking

IBIS trunking is a method which is finding favour for common base station applications. All channels are used for communication with no dedicated calling channel, thus making the most effective use of the channels available.

On a CBS system all users access the base station, which operates talk-through, by radio. Each user is allocated his own continuously coded squelch system (CTCSS) tone to access the base station and to prevent overhearing by other users. A trunked common base station contains the transmitters and receivers for a number of radio channels and when the system is quiescent transmits bursts of an in-band (speech band) free-tone – 1953 Hz is one that is used – on one channel only. The mobiles, which are fitted with decoders to recognize both the free-tone and their CTCSS tone, scan all channels looking for one containing free-tone onto which they lock. Should the mobile operator press the transmit button the transmitter will only be enabled when the mobile receiver is on a channel containing either the free-tone or its CTCSS code. As soon as the channel becomes occupied the base station commences to radiate free-tone on another channel and mobiles of other user groups re-lock onto the new free channel. When either a mobile or the control station, which is a fixed mobile, makes a call it is on the channel radiating free-tone and the transmission contains the CTCSS code allocated to that user group. The repeater re-transmits the CTCSS code which activates the calling mobile and all other mobiles within that group. Mobiles of other user groups sensing the alien CTCSS tone leave the channel and continue scanning for a free channel. If all channels are engaged, the caller hears a 'busy' tone and must try again later.

17.4 Trunking to MPT 1327 specification

Specification MPT 1327 produced by the Radiocommunications Agency lays down a signalling standard for trunked private land mobile radio systems. The system uses fast frequency shift keying (FFSK) at 1200 bit/s with time division random access. The mobiles operate two-frequency semi-duplex and the base stations or trunking system controller (TSC) operate full duplex. A calling channel is used for call establishment but may be either dedicated or nondedicated. After a call is established messages are transmitted over a traffic (working) channel. The basic components of the signalling format, in sequence, are:

1. The link establishment time (LET), a period of transmission of undefined modulation.
2. A preamble comprising a sequence of bit reversals 1010. . .10, minimum 16 bits and ending with a binary zero for the receiver data demodulator to achieve synchronization.
3. A message. A contiguous transmission of a synchronization sequence, address codeword and, where appropriate, data codewords.
4. A hangover bit, H, of either binary zero or binary one, added to the last transmitted message identifying the end of signalling transmission.

The message synchronization sequence for both control and message channels consists of 16 bits, the first bit being a binary one on a control channel and binary zero on a traffic channel.

Messages, whether address or data, are transmitted in 64 bit codewords. The first bit, bit 1, indicates the status of the codeword. Binary one denotes an address word, binary zero, a message word. Bits 2–48 are the information field and bits 49–64 are check bits.

References

Macario, R.C.V. (1991). *Personal and Mobile Radio Systems*. Peter Peregrinus, London.

Parsons, J.D. and Gardiner, J.G. (1989). *Mobile Communication Systems*. Blackie, London and Glasgow.

Radiocommunications Agency (1991). MPT 1327, *A Signalling Standard for Trunked Private Land Mobile Radio Systems*.

Radiocommunications Agency (1993). MPT 1318, *Engineering Memorandum, Trunked Systems in the Land Mobile Radio Service*.

Winder, S.W. (2001). *Newnes Telecommunications Pocket Book*. Butterworth-Heinemann, Oxford.

18 Mobile radio systems

18.1 Paging

Where an immediate reply to a message is not essential, paging systems, which are basically a one-way path from controller to mobile, are an economic option. Paging is also spectrally economic, only one frequency being required and messages being of very short duration. The recipient of a call may be merely alerted by means of a tone emitted by the paging receiver or, if silent alert is required, a vibration. Alternatively, pagers displaying, and storing, alpha-numeric messages are readily available. Message handling services are also offered on some paging systems.

On-site paging systems are usually owned or rented by the user and consist of either a radio base station or, in the case of a non-radio system, an inductive loop around the perimeter of the service area. The details of on-site paging systems and the Radiocommunications Agency specifications governing them are:

RF frequencies:

on-site, 26 MHz and 31 MHz bands, MPT 1365
49.0–49.5 MHz, MPT 1335
160–161 MHz
458–459 MHz, MPT 1305

Channel spacings:

12.5 kHz in the VHF bands and 25 kHz in UHF band

Modulation:

FM, FFSK data

Wide area, national or international systems are generally owned by a network provider to whom the user subscribes for the service. There are frequencies allocated in the high VHF and UHF bands. The appropriate specifications are MPT 1308 for receivers and MPT 1325 for transmitters.

The Post Office Code Standardization Group (POCSAG) standard which has been adopted by the CCITT allows the transmission of messages at a rate of 10 calls per second and a capacity of 1 million pagers with up to four alternative addresses per pager. The recommended bit speed is 512 bit/s with direct FSK deviating the transmitter frequency by ± 4.5 kHz. Positive deviation indicates a binary zero and negative deviation a binary one.

In America the Golay sequential code is widely used for paging addressing. The format for tone-only paging consists of 12 address bits followed by 11 parity bits for each 23 bit word. Two bit rates are used, 300 bit/s for address codewords and 600 bit/s for message codewords.

There are European paging and message systems in operation including Euromessage and the frequency agile European Radio Message Service (ERMES).

18.2 Cordless telephones

CT1

The first generation of cordless telephone, still very popular in the UK, is designed to serve the domestic environment with a range of 100 m. The base station transmits on one of eight frequencies between 1.6 MHz and 1.8 MHz, and the handset on one of eight frequencies in the 47 MHz band. Frequency modulation is used with a deviation of ± 4 kHz at the base station and 2.5 kHz at the handset. MPT 1322 applies and permits an erp of 10 mW at both the base station and the handset.

CT2

The second generation of cordless telephone. The services, under various names, telepoint, phonezone, etc. were planned to provide for the general public a lower cost alternative to the cellular radio-telephone networks which were seen at the time as a businessman's preserve. However, the low market uptake caused their demise in the UK although services are thriving in some European and Far Eastern countries.

One hand-portable transmitter/receiver can operate to a local base station installed in the home or office, or through one of a number of multi-channel base stations with a range of approximately 200 m installed in public places. While away from the local base station the subscriber must initiate a call: calls cannot be made from the PSTN to a CT2 subscriber's handset.

The operational frequencies are in the band 864.1 to 868.1 MHz and employ time division multiple access. Speech is digitized at 32 kbit/s, stored, and then transmitted at 64 kbit/s in 1 ms slots. This leaves the alternate 1 ms slots available for the digitized and stored speech of a reply. Duplex operation is achieved in this way on a single radio frequency.

Inter-operability between networks was to be ensured by specification MPT 1375 *Common Air Interface*.

Digital European cordless telephone (DECT)

A pan-European system complying with an ETSI standard, DECT operates in the 1880 to 1900 MHz band. It offers data handling facilities and the ability for a subscriber to receive calls while away from the local base station. The techniques are similar to those used for GSM although, because the mobile is virtually stationary, the constraints on data transmission are less severe and no hand-off is required. The 20 MHz RF bandwidth is divided into 13 carriers spaced at 1.7 MHz intervals, each carrier containing 12 TDMA channels with GMSK modulation.

CT3 is developed from DECT and operates in the band 800 to 900 MHz. Each 8 MHz section of that band is divided into 1 MHz blocks, each containing sixteen 1 ms time slots.

18.3 Trunked radio

Bank III trunked radio

So-called because it occupies part of the now redundant television band III, 174 to 225 MHz, and often referred to as private mobile radio, this is a subscriber access system. The network has almost nationwide coverage in the UK, arranged on both local and regional bases. The network permits radio communication between a fixed office station and mobiles, mobile to mobile, and limited access from mobiles to the PSTN.

The relevant Radiocommunications Agency specifications are:

MPT 1323	*Angle-modulated Radio Equipment for Use at Base and Mobile Stations Private Mobile Radio Service Operating in the Frequency Band 174–225 MHz.*
MPT 1327	*A Signalling Standard for Trunked Private Land Mobile Radio Systems.*
MPT 1343	*System Interface Specification for Radio Units to be Used with Commercial Trunked Networks Operating in Band III sub-band 2.*
MPT 1347	*Radio Interference Specification for Commercial Trunked Networks Operating in Band III sub-band 2.*

MPT 1352   *Test Schedule. For Approval of Radio Units to be Used with Commercial Trunked Networks Operating in Band III sub-band 2.*

Frequency modulation is used with FFSK signalling, a channel spacing of 12.5 kHz and TDMA techniques (Section 17.4).

Trans-European trunked radio (TETRA)

A new European standard for digital trunking. The system is aimed at PMR users and will employ TDMA with four users sharing a 25 kHz radio channel with the option of two users on a 12.5 kHz channel. Speech will be digitally encoded at 4.8 kbit/s and transmitted at 36 kbit/s (gross bit rate including error correction overhead) using $\pi/4$ differential quaternary phase shift keying (DQPSK).

Digital short range radio (DSRR)

This utilizes trunking techniques but with no infra-structure of control or switching centres; mobiles work to mobiles via a locally installed repeater. It offers great potential for short-term hire for local events. The band 933 to 935 MHz has been allocated to the service and accommodates 76 traffic and 2 control channels separated by 25 kHz. The operating procedure may be either single-frequency simplex or two-frequency semi-duplex. Mobiles remain on standby waiting to receive their selective signalling code (SSC) on one of the two control channels; mobiles with even serial numbers normally listen on control channel 26 and those with odd serial numbers on control channel 27. A mobile wishing to initiate a call first scans for a free traffic channel and, having obtained one, selectively addresses the intended recipients of the call over the relevant control channel. The recipient mobiles are then instructed automatically by the system to go to the appropriate traffic channel.

The modulation is GMSK at 4 kbit/s for addresses and 16 kbit/s for speech and data messages. The maximum transmitter erp is 4 W.

18.4 Analogue cellular radio-telephone networks

Analogue networks are becoming obsolete and are being replaced by digital cellular networks. There are no analogue networks operating in the UK and the frequency bands that they used to occupy are now reassigned to GSM operators.

However some countries still have an analogue service based on AMPS or TACS. These provide national coverage with roaming facility and hand-off from cell to cell as a mobile travels across the network, calling for more complex signalling than local trunking.

The control system must know at all times the location of all operational mobiles. It does this by continuously monitoring signal strength and instructing the mobile to change channel as it crosses cell boundaries.

To permit the maximum re-use of frequencies mobiles are instructed by the system controller to reduce transmitted power to the minimum necessary for acceptable communications in either eight or five steps, depending on the mobile classification, to 0.01 W. Mobiles are classified by their maximum effective radiated power output:

Class 1	10 W erp, vehicle mounted
Class 2	4 W erp, transportable
Class 3	1.6 W erp, transportable
Class 4	0.6 W erp, hand-portable

The channel separation is 25 kHz with a peak speech deviation of ± 9.5 kHz. Signalling is by Manchester encoded PSK at 8 kbit/s with 6.4 kHz deviation.

To discriminate between stations transmitting on the same RF channel an FM supervisory audio tone (SAT) is transmitted concurrently with speech at ± 1.7 kHz deviation. There are three such tones: 5970, 6000 and 6030 Hz.

An FM signalling tone (ST) of 8 kHz transmitted for 1.8 seconds (deviation = ± 6.4 kHz) indicates the clear-down of a conventional telephone call.

The advanced mobile phone service (AMPS) is the American cellular system designed for 30 kHz channel spacing with a 10 kbit/s signalling rate. TACS is based on the AMPS system but modified for reduced channel spacing. Narrower channel spacing systems (NAMPS and NTACS) are also found in some countries.

18.5 Global system mobile

Global system mobile (GSM), formerly known as Groupe Speciale Mobile, is the pan-European digital cellular radio-telephone service. The operational requirements for GSM are severe, e.g. it must operate satisfactorily to a person walking, or in a slowly moving vehicle, in a street where much of the furnishings will introduce multi-path fading, and operate to a train travelling at 250 km/hr where Doppler frequency shift becomes significant. To reduce the corruption, a high degree of error detection and correction must be applied which increases the occupied bandwidth. To compensate, to some extent, for this, the system takes full advantage of the redundancy in speech to reduce the bandwidth during synthesis.

GSM operates full duplex in the band 890–915 MHz, up-link and 935–960 MHz, down-link.

A combination of FDMA and TDMA is employed. Each allocated band of 25 MHz is divided into 125 carriers spaced 200 kHz apart. The subdivision of each transmitted bit stream into 8 TDM time slots of 540 μs gives 8 channels per carrier and 1000 channels overall. Modulation is Gaussian minimum shift keying (GMSK).

Speech synthesis is by speech codec using linear predictive coding. It produces toll quality speech at 13 kbit/s.

18.5.1 Emergency locator service

The FCC in the US has placed a requirement on mobile phone operators to provide an emergency location service, referred to as E911. The aim is to locate people who need emergency services, but are not able to indicate their location either through injury or through lack of local knowledge. The locator service may also deter hoax calls (or help to catch people making hoax calls) and could have commercial applications in providing directions to businesses such as garages and restaurants.

There are two options: handset based solutions or network-based solutions. The handset-based solutions generally rely on signals picked up from the global positioning system (GPS). Network-based solutions rely on the timing difference in the arrival of signals from neighbouring cell sites. In network based solutions, the E911 requirement is for the mobile phone to be located within 100 metres at least 67% of the time and to within 300 metres in 95% of cases. In handset-based solutions, the location distances are 50 metres and 100 metres, respectively.

18.5.2 Data over GSM

Standard GSM data service allows a maximum of 9600 baud data rate transmission. An enhanced data rate of 14.4 kbps is also possible with some terminals.

The data link can be either synchronous or asynchronous and the data is conveyed through the network using an ISDN-based data protocol, the ITU-T standard V.110 rate adaptation scheme. Asynchronous transmission requires start and stop bits to be added, so the net data rate is reduced to 7680 bits per second (bps). Synchronous transmission potentially allows a faster data rate, but error checking is employed that requests re-transmission of data when errors are identified and this can produce a lower overall data throughput.

18.5.3 High-speed circuit switched data (HSCSD)

HSCSD allows higher data rates by combining up to four 14.4 kbps data channels together. A maximum data rate of 57.6 kbps is possible.

18.5.4 General packet radio service (GPRS)

GPRS combines up to eight 14.4 kbps channels to give data rates of up to 115.2 kbps. At these rates, Internet browsing is possible. GPRS supports X.25 and IP based transmission, which includes wireless application protocol (WAP) data.

18.6 Other digital mobile systems

Second generation mobile systems are digital systems. The GSM system has been described in Section 18.5. This section describes the GSM1800 system, the personal communications system (PCS1900, sometimes known as personal communications network or PCN) and third generation systems.

18.6.1 GSM1800 (DCS1800)

The GSM1800 system operates in the 1840–1880 MHz band, with a carrier spacing of 200 kHz. Each carrier gives multiple access through time division multiple access (TDMA). It is a lower power version of GSM900, which operates in the 900 MHz band. A GSM1800 handset has a maximum power output level of 1 watt, rather than the 2 watts of a GSM900 handset.

Voice coding can be either: half-rate using VSELP at 5.6 kbps, full-rate using RPE-LTP at 13 kbps, or enhanced full rate (EFR) using ACELP at 12.2 kbps. Convolutional coding is used at a rate of $\frac{1}{2}$ to reduce the effects of noise and errors on the radio channel. The radio carrier modulation is Gaussian minimum shift keying (GMSK).

18.6.2 PCS1900 (IS-136, or PCN)

The personal communications system (PCS) is based on GSM and used time division multiplexing to provide separate channels over each radio frequency. Convolutional coding is used at a rate of $\frac{1}{2}$ to reduce the effects of noise and errors on the radio channel. Each carrier is spaced 30 kHz apart and offset quadrature phase-shift keying (OQPSK) is used to modulate the carrier.

The voice-coder (vocoder) used by PCS1900 is either full-rate using VSELP at 7.95 kbps or enhanced full rate (EFR) that uses ACELP at 7.4 kbps. The EFR scheme was developed for use in PCS1900 and has since been adopted by both GSM900 and GSM1800 networks because it improves the audio quality. Previously, GSM used either a half-rate (HR) or full-rate (FR) vocoder.

18.6.3 Third generation systems

The so-called 3G systems have been described using a number of terms: UMTS, IMT-2000, cdma-2000, w-cdma, etc. There are variations between the proposed standards, but in general they all use code division multiple access (CDMA), convolutional coding, quadrature phase shift keying and time or frequency division duplexing.

Third generation systems will operate in the 1885–2025 MHz band and in the 2110–2200 MHz band. Initially the lower band between 1980 and 2010 MHz will not be allocated, nor will the upper band between 2170 and 2200 MHz. However, the full bands will be available when existing services in these bands can be moved (such as PCS1900, PHS and DECT).

The code division multiple access (CDMA) modulation scheme allows several channels to occupy the same frequency band. CDMA is a form of spread spectrum encoding. The digitized speech signals are coded using a unique code, known as a spreading code, so that one bit of speech code produces many encoded bits. When demodulated and mixed with the same spreading code, the original data is recovered. Receivers that use a different spreading code are unable to recover this data, but they are able to recover data intended for them.

18.7 Private mobile radio (PMR)

These systems are privately owned or rented and may be used only by the owner's own staff for the purpose of the owner's business. The leasing of air-time is forbidden. Common base station or community repeater systems where several users subscribe to, and use, the station is often also referred to as PMR.

18.8 UK CB radio

27 MHz band: 27.60125 to 27.99125 MHz

40 channels at 10 kHz spacing.
Max. erp, 2 W; max. transmitter output power, 4 W.
Antenna: single rod or wire, 1.5 m overall length, base loaded.
If mounted higher than 7 m transmitter output to be reduced at least 10 dB.
Modulation: FM only, deviation ±2.5 kHz max.

934 MHz band: 934.025 to 934.975 MHz

20 channels at 50 kHz spacing (may be reduced later to 25 kHz).
If synthesizer used spacing may be 25 kHz on precise channel frequencies specified.
Max. erp, 25 W; max. transmitter output power 8 W. If antenna integral, max. erp, 3 W.
Antenna: may have up to four elements, none exceeding 17 cm.
If mounted higher than 10 m, transmitter output to be reduced at least 10 dB.
Modulation: FM only, deviation ±5.0 kHz max.

Spurious emissions: for both bands, not exceeding 0.25 μW except for specified frequency bands where the limit is 50 n W.

For latest full specifications see publications MPT 1320 (27 MHz band) and MPT 1321 (934 MHz band) obtainable from the Information and Library Service of the Radiocommunications Agency.

References

Belcher, R. *et al.* (1989). *Newnes Mobile Radio Servicing Handbook*. Butter-worth-Heinemann, Oxford.
Macario, R.C.V. (1991). *Personal and Mobile Radio Systems*. Peter Peregrinus, London.
Winder, S.W. (2001). *Newnes Telecommunications Pocket Book*. Butterworth-Heinemann, Oxford.

19 Base station site management

The information in this chapter applies mainly to sites for stations operating at VHF and above. Some conditions will be different for stations working at lower frequencies due to the generally physically larger transmitters and antennas.

19.1 Base station objectives

The objectives of a radio base station are:

1. To provide an adequate service over the whole of the desired service area in a well-defined manner. Special consideration must be given to those areas where, because of their location, weak signals or a high ambient electrical noise level might be expected.
2. To create the minimum interference with other radio users and be designed to receive the minimum from other users.
3. To be electromagnetically compatible with neighbouring installations, possibly on the same site.
4. To create the minimum impact on the environment.

19.2 Site ownership or accommodation rental?

The question of whether to privately develop a base station site or rent space on someone else's site may be resolved by the planning authorities rather than the developer. Before planning consent will be given the developer must normally prove that no existing site will meet the requirements. However, assuming that land is available and planning consent might be forthcoming, other factors to consider are costs, both initial and recurring, and access for services, materials during building and tower erection and for maintenance during bad weather (*Table 19.1*).

19.3 Choice of site

The first stage in selecting a site is to look for an existing site which might appear to cover the service area. Information may be available from the operators of such sites regarding their coverage, and coverage tests may be made or computer predictions obtained. For a

Table 19.1 Factors affecting the decision to own or rent

	Ownership	Rental
Initial costs	Legal charges	Legal charges
	Land clearance	Connection fee
	Provision of services	
	Erection of buildings and mast or tower	
	Installation of equipment	Installation of equipment
	Provision of standby power	Provision of standby power
Recurring costs	Maintenance of site	Rental
	Power	Normally included in rental
	Insurance (other than radio equipment)	Normally included in rental
Other considerations	Very long commissioning time	Very short commissioning time
	Site under user's control	Site under landlord's control
	Space may be leased to other users creating revenue	Accommodation will be shared with other users (install equipment in locked cabinet to preserve security)

receiving site the ambient noise level at the base station must also be considered. It should be realized that a site, or an antenna, can be too high, particularly when coverage is required at very short range or if interference from other users is likely to occur. Antennas radiate the minimum signal along the line of their elements so that vertically polarized antennas, i.e. with the E plane vertical, provide very little, if any, signal immediately below them. This creates a hole in the coverage, 'the Polo mint effect', close to the antenna.

Where there is capacity on an existing site the possibilities are:

1. Sharing an existing lightly used radio channel, e.g. common base station.
2. Sharing equipment space in an existing building or providing own building or extension on a communal site.
3. Sharing an antenna by means of multi-coupling or erecting own antenna on existing structure.

19.4 Masts and towers

Apart from short, unguyed poles mounted on the side of buildings, antenna supports fall into two types: guyed masts and self-supporting towers.

A guyed mast may be a single pole, or a square or triangular section lattice structure. The range of single-pole masts includes those of about 4 m maximum height mounted on chimney stacks and secured by wire lashings, and poles supported on the ground or the roof of a building. The latter types may be up to 30 m high. They are usually constructed from steel or aluminium hollow tubular sections – the bottom section is invariably steel – with a set of guys at about every 5 m. Mast erection is specialist work but to estimate the approximate height that a site can accommodate, a four-guy, single-pole mast requires a minimum spacing for the anchorage points in the form of a square with sides equal to half the height of the mast. The guys may be of steel, possibly plastic-coated, or more likely of synthetic fibre. Steel guys may affect the directivity of antennas in close proximity, and if the guys corrode at their fixing points, intermodulation interference may result.

Self-supporting towers come in all heights up to many hundreds of feet and may be made of a variety of materials from steel sections to concrete.

The factors determining the type of structure to be erected are the number and types of antenna, and the site conditions.

Antenna considerations:

- Physical dimensions and space requirements.
- Weight.
- Wind loading.
- Directivity. For example, the beam width of a 6.5 GHz, 3.0 m dia. microwave dish is 0.9 degrees. A tower supporting such an antenna must twist, therefore, less than about 0.3 degrees in the highest wind likely to be experienced at the site.
- Access for riggers to antenna mounting positions.

Site conditions:

- Availability of space for tower footings, and guys and anchorages if a mast is contemplated.
- Stability and type of ground.
- Weather conditions. High winds. Accumulation of ice and snow. Build-up of ice increases wind loading of antenna – and of mast or tower – considerably. There is no point in using grid microwave antennas where icing is likely to occur.
- Aesthetic and planning consent considerations.

19.5 Installation of electronic equipment

For radio equipment, safety is one of the first priorities and in addition to the electrical considerations, physical hazards must be avoided.

Adequate working space must be provided around the equipment racks. Much equipment is now wall-mounted but tall, floor-standing cabinets must be firmly bolted down to prevent toppling when units are partially withdrawn for servicing and cables need to be routed safely. Overhead cable trays are generally considered the most satisfactory method but, to avoid interaction, it is recommended that cables for the various functions be segregated to reduce cross-talk and interference. Where cabling enters a cabinet at floor level, mounting the cabinet on a hollow plinth and running the cables through the plinth avoids damage to the cables and persons tripping over them.

For low voltage installations (see classifications below) the IEE wiring regulations apply. For high voltage installations, recommendations are given in BS 6701: part 1: 1990 which also covers accommodation, lighting levels and access arrangements.

Classification of installations:

- *Extra low voltage.* Normally not exceeding 50 V AC or 120 V DC.
- *Low voltage.* Exceeding 50 V AC or 120 V DC but not exceeding 1000 V AC or 1500 V DC between conductors, or 600 V AC or 900 V DC between conductors and earth.
- *High voltage.* Exceeding 1000 V AC or 1500 V DC between conductors, or 600 V AC or 900 V DC between conductors and earth.

19.6 Earthing and protection against lightning

Recommendations for earthing are given in BS 7430: 1991. This document covers the earthing of equipment and the principles of earthing for protection against lightning. All equipment metalwork must be bonded together and connected to the electricity supply earth point, the main earth terminal. In addition a connection must be made to the earthing system provided for protection against the effects of lightning.

Antenna systems by their nature are vulnerable to lightning strikes. Nearby taller structures may reduce the risk but precautions must still be taken. The zone of protection – a cone with its apex at the tip of the protecting structure and its base on the ground forming an angle of 45° to the perpendicular – does not necessarily protect structures above 20 m high. BS 6651: 1985 is the *Code of Practice for the Protection of Structures against Lightning.* Lightning protection begins at the top of the mast or tower where, ideally, the highest item should be a finial bonded to the tower. However, a finial mounted alongside an omni-directional antenna will affect is radiation pattern. Grounded antennas are preferred and the outer conductor of each feeder cable must be

218

bonded to the tower at the top and bottom of its vertical run. Grounding kits for the purpose are obtainable. The feeder cables must also be bonded to an earthing bar at their point of entry to the equipment building. The codes of practice should be consulted regarding the routes to be taken by earthing tape and the methods of jointing. An important point is that neither very sharp bends nor 'U' bends are created.

The legs of the tower must be bonded together with earthing tape and each leg connected to an earthing system which may consist of a buried plate or rods. Several rods may be necessary for each subsystem to attain the specified earth resistance of not more than 10 ohms. The rods should, except in rock, be driven into ground which is not made up or backfilled or likely to dry out. Where several rods are necessary to achieve the specified earth resistance these should be spaced well apart, the reduction in earth resistance being small with parallel rods closely spaced. Joints to facilitate testing of each subsystem separately must be provided either above ground or in a purpose-built pit or chamber. *Figure 19.1* shows a typical earthing system.

Figure 19.1 Typical example of good earthing practice (© Crown Copyright 1991, Radiocommunications Agency)

The system should be tested regularly, at least annually, and the results recorded in a lightning protection system log book. The test comprises:

1. Measurement of the resistance to earth of each termination network and each earth electrode.
2. Visual inspection of all conductors, bonds and joints or their measured electrical continuity.

BS 7430 recommends the method of testing.

Other vulnerable systems are overhead cables for either the electricity supply or telephone and control purposes.

19.7 Erection of antennas

The installation of the antenna probably influences the performance of a radio system more than any other part. The heights involved render supervision of the installation difficult and, although antennas are relatively inexpensive, labour charges raise the cost of their replacement, and bad weather may delay rigging work. It is therefore important that good quality antennas are specified and that the initial installation is correct. In addition to the physical mounting arrangements, the dispositions of antennas on the structure are important. Some aspects are detailed below and MPT 1331 contains more information.

19.7.1 Directivity

The radiation pattern diagrams provided in the antenna manufacturers' catalogues are produced by an isolated antenna. They can be greatly affected by the proximity of the support structure, e.g. if a dipole is mounted close to the leg of a tower, the leg behaves as a reflector and the radiation pattern of the antenna approaches that of two-element Yagi. The less directional the antenna, the worse the effect. Omni-directional antennas should not be mounted on the side of a mast or tower if it can be avoided. If an omni-directional antenna must be so mounted it should be spaced two wavelengths from it. This is difficult to achieve at the lower frequencies considering the antenna dimensions, weight and windloading. The rear element of a Yagi antenna must be one wavelength clear of the structure. To obtain a nominally omni-directional pattern a better, albeit expensive, solution is to mount a number of antennas, phased correctly, around the tower.

All antennas should point directly away from the structure, i.e. at right angles to the side of the tower.

The proximity of metalwork also adversely affects the VSWR of an antenna, causing standing waves with their additional voltage stresses on the feeder and possibly producing interference.

19.7.2 Practical aspects of antenna installation

All installation components must be made from compatible materials, electrically dissimilar metals, e.g. steel and brass, brass and aluminium, being avoided at all costs; they will corrode and cause intermodulation. Preferred materials are aluminium, and galvanized or stainless steel. Mounting brackets must be secure – remembering that antennas and feeders vibrate – but not overtightened so as to distort and weaken antenna support booms.

Connectors of the correct impedance, and preferably with a captive centre pin, are vital. Type N connectors are available at either 50 or 70 ohms impedance, are robust and can be used at frequencies up to 10 GHz. UHF 83 series connectors, although very robust, are of imprecise impedance and, from first-hand experience of interference aggravated by a UHF 83 used to connect an antenna feeder to a receiver tail, should not be used on base station installations. Assembly instructions for a range of connectors are shown in Chapter 23. All joints must be waterproofed, preferably by first wrapping with a self-amalgamating tape for two inches either side of the joint, then a layer of PVC tape covered by a further layer of Sylglass or Denso tape. When installing feeder cables the required length should be loosely coiled and taken up the tower and paid out as it is secured, working from top to bottom. It must not be dragged from a drum on the ground. Feeders must be cleated at the intervals specified by the manufacturer. Route cables where they are easily accessible but least likely to suffer physical damage.

Health and safety

Apart from the physical risks to riggers, there are radiation hazards. Large amounts of RF power may be radiated from antenna systems and research is continuing into its effects on health. The most recent recommendations of the National Radiological Protection Board and the Department of Trade and Industry should be sought and followed. Riggers should be aware of the power levels present on a structure before climbing, and equipment switched off where levels are considered unsafe.

19.7.3 Antenna checking and fault finding

The majority of VHF and UHF antennas present a short-circuit across the antenna and to ground as a means of reducing their vulnerability to static charges. An ohmmeter applied to the lower end of the feeder will indicate a circuit through the feeder and antenna but may also indicate a short-circuit or low resistance fault, water in the feeder for instance. Most antenna system faults increase the VSWR present on the feeder cable. A good antenna, operating within its design bandwidth, exhibits a VSWR of 1:1. When connected to a length of co-axial cable this may be raised and a VSWR of 1.5:1 for the system is generally considered acceptable. However, faults can be masked by conditions occurring on the feeder, and a measurement taken with a standing wave, or forward and reflected power meter at the bottom of the feeder, may not indicate the true condition of the antenna. A measurement taken with an accurate dummy resistive load connected to the top of the feeder will prove the cable, and a measurement of the VSWR should also be taken directly at the antenna. A record of the VSWR at installation and any subsequent measurements is helpful.

19.8 Interference

There are two sources of radio interference induced by antenna systems. One occurs from strong signals radiated by either the system's own transmitter antenna or from a co-sited system operating on a close neighbouring frequency. The other source is intermodulation, 'rusty bolt effect', the mixing of two or more signals to produce the interfering frequency.

19.8.1 Antenna isolation

One solution to the problem of direct radiation is to space the antennas so that there is sufficient isolation between them. Because minimum radiation is present immediately below and above the ends of the elements of vertically polarized antennas, maximum isolation occurs when such antennas are mounted in a vertical line. Maximum radiation, and hence minimum isolation, occurs when antennas are mounted side-by-side. The degree of isolation depends on the spacing but a figure of 40–45 dB between a transmitter and a receiver antenna, and 20–25 dB between transmitting antennas, should be the target. To achieve 25 dB isolation requires a vertical separation of 0.9 wavelengths between the centres of vertically polarized dipoles; 45 dB requires a vertical separation of 3 wavelengths. The spacing for horizontal separation of vertically polarized antennas is 2.5 wavelengths for 25 dB isolation and 25 wavelengths for 45 dB. To preserve this isolation the feeder cables should also be separated over their routes.

Antenna multi-coupling

A precise method of controlling the RF isolation between mutually sited systems where the frequencies are moderately spaced is for several systems to share one antenna by using multi-coupling techniques (see Section 19.9).

19.8.2 Intermodulation

Any two or more RF signals applied to a non-linear device intermodulate, that is, they combine to form additional frequencies. *Table 19.2* lists the combinations for two input frequencies. It must be remembered that the side frequencies produced by modulation of the original carriers will also be present.

Table 19.2 Low order intermodulation products

2nd	$A + B$	5th	$3A + 2B$	7th	$4A + 3B$
	$A - B$		$3B + 2A$		$4B + 3A$
			$3A - 2B^*$		$4A - 3B^*$
3rd	$2A + B$		$3B - 2A^*$		$4B - 3A^*$
	$2A - B^*$		$4B + A$		$5A + 2B$
	$2B + A$		$4A + B$		$5B + 2A$
	$2B - A^*$		$4A - B$		$5A - 2B$
			$4B - A$		$5B - 2A$
4th	$2A + 2B$				$6A + B$
	$2A - 2B$	6th	$5A + B$		$6B + A$
	$2B + 2A$		$5B + A$		$6A - B$
	$2B - 2A$		$5A - B$		$6B - A$
	$3A + B$		$5B - A$		
	$3A - B$		$4B + 2A$		
	$3B + A$		$4B + 2B$		
	$3B - A$		$4B - 2A$		
			$4B - 2B$		
			$3A + 3B$		
			$3A - 3B$		
			$3B - 3A$		

*indicates in-band products. No eighth-order products fall in-band but ninth-order in-band products are produced by:

$5A - 4B$

$5B - 4A$

19.8.3 Control of intermodulation

A clue to the source of interference will be obtained from examination of the relationship of all frequencies produced on site, and also those produced nearby. That the side frequencies produced by modulation must be considered was proved in an intermodulation situation where the interference was only received when a nearby band III television transmitter radiated a peak white signal.

Corroded joints in metalwork close to the site, receiver RF stages subjected to very large signals and transmitter output stages are all non-linear devices and can create intermodulation.

Regular maintenance of the metalwork around the site reduces the chances of intermodulation from this source, but it is difficult to see and eliminate all corrosion on an inspection. Should interference from this cause be suspected after elimination of the other sources, a successful method of locating the offending joint requires a receiver tuned to the interfering signal and fitted with a whip antenna. While receiving the interference, a very sharp null in signal strength occurs when the tip of the whip points directly at the source of the signal.

Receiver RF amplifiers are usually designed to cope with small signals. When a strong signal radiated from a nearby transmitter is received the stage may well be overloaded and driven into non-linearity, or even into a blocked state where the receiver is effectively dead and may take some seconds to recover after the signal is removed. In the non-linear state the receiver is in an ideal condition to create intermodulation. If this situation is suspected a method of identifying it is to install a variable RF attenuator in the antenna feed to the suspect receiver. With zero attenuation the interference will be received but, if the receiver is the cause, increasing attenuation will produce little reduction of the interference until a point is reached where the interference suddenly disappears – the receiver at this point has re-entered a linear state. If the interference reduces gradually to zero with increased attenuation, the source is elsewhere.

Most amplifiers are designed to be linear, that is, the output signal level will follow that of the input signal. However, with a sufficiently high input signal overloading occurs, the amplifier becomes non-linear and compression of the signal results (*Figure 19.2*). The point where 1 dB of compression occurs is a commonly referred to amplifier parameter. If the input level is increased further the gain of the amplifier is reduced until saturation is reached when the output level can no longer increase.

At the onset of non-linearity harmonics and intermodulation, if any other frequencies are present, are produced. The strength of these rises rapidly with an increase of input, and the point where an extension of the almost linear portion of their curve crosses an extension of the linear gain line of the amplifier is the third-order intercept – usually it is only the third-order products which are considered for this purpose.

224

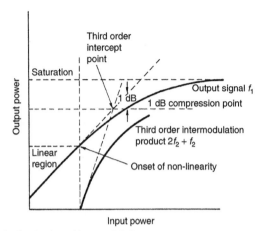

Figure 19.2 Production of intermodulation

The class C operated output stages of transmitters are by design non-linear. Strong signals from a neighbouring transmitter or antenna applied via the antenna feed will mix with the transmitted frequency to create intermodulation products. Increased isolation between antennas and feeders may be the simplest remedy. Alternatively, a circulator or filters may be connected to the output of the transmitter,

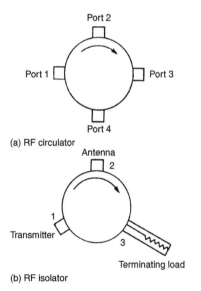

Figure 19.3 RF circulator and isolator

and the possibility of direct radiation from transmitters must not be ignored.

A circulator is a uni-directional device with either three or four ports as shown in *Figure 19.3(a)*. A signal entering at port 1 will leave at port 2 but not at ports 3 or 4. Similarly, a signal entering at port 2 will leave at port 3 but not ports 1 or 4. Circulators are used for combining the outputs of transmitters for application to a single antenna but if a three-port circulator is connected as in *Figure 19.3(b)* it forms an isolator. Typical isolation over a 1.5% bandwidth is 20–30 dB with an insertion loss of 0.7 dB.

19.9 Antenna multi-coupling

The sharing of antennas and feeders between a number of co-sited services is not only a precise method of controlling frequency isolation, it is an economic solution to the problems of antenna and tower management. An antenna may be shared between a number of receivers or transmitters, or shared simultaneously by both receivers and transmitters. Receiver sharing requires a splitter, i.e. a filter, to separate the frequencies for each receiver and an amplifier to make up the filter losses. Transmitters can share an antenna through circulator/isolators and filters. To obtain the required selectivity, stability

Figure 19.4 Typical sub-band transmitter/receiver sharing (© Crown Copyright 1991, Radiocommunications Agency)

and power handling capability the filters are often solidly constructed cavity resonators with an insertion loss of 1 dB and a bandwidth of 0.1% to the −20 dB points. *Figure 19.4* shows the layout of a transmitter/receiver sharing system.

19.10 Emergency power supplies

The installation of equipment to provide power for all the equipment on a site during a failure of the public supplies has declined in recent years. But although lengthy power failures seem to be a thing of the past, they could recur, and electricity supplies do still fail for short periods. To provide a complete supply for large station necessitates a standby generator. On smaller stations batteries may be capable of supplying the demands of the electronic equipment and emergency lighting for one or two hours. The trend on multi-user shared sites appears to be for the individual user to provide the emergency power, derived from batteries, for his own equipment. Factors to consider are:

1. Standby motor generators are expensive. They need housing, ventilation and fuel storage. They take time to start up, 45 seconds being typical.
2. Uninterruptible power supplies (UPS). Where no interruptions can be tolerated the most efficient supply is provided by batteries alone but the equipment must be designed to operate directly from a DC, low voltage supply.

   To supply power at 220/240 V AC an inverter is the accepted method. Double inverters permanently charge a battery from the AC supply and the battery drives an oscillator to reproduce the AC supply voltage, a very inefficient process. Also, should any item of the inverter break down the supply is lost. This is overcome in some inverters by a switch which transfers the load to the public supply when a failure of the inverter occurs. Single inverters charge a battery and supply the equipment directly from the public supply until a failure occurs. When this happens the supply changes over to the battery and oscillator. Single inverters are more efficient but there is a short break in supply while the change-over occurs. Adequate stabilization and filtering to avoid disturbances on the supply must be provided. One possible hazard with all inverter supplies is that where more than one inverter is used, the outputs of the inverters are unlikely to be in phase and a 480 V difference could exist between the supplies for two items of equipment.

For all battery-operated supplies correct charging of the battery is crucial. Nickel–cadmium batteries require constant-current charging with a fall to the trickle rate when fully charged and sophisticated circuitry is needed to identify the precise point of full charge (see Section 21.3.2).

Lead acid batteries produce hydrogen if over-charged so their storage must be designed so that no danger, either to personnel or of explosion, exists. For reliability, natural ventilation is preferred to forced. Recommendations for the accommodation of lead acid batteries other than sealed are given in BS 6133: 1985. Batteries which gas also need more frequent maintenance. Because of the reduction in performance of batteries at low temperatures the ambient temperature of the battery storage room should not fall below 4°C, and to reduce water evaporation lead acid batteries should not be run continuously above 40°C.

The period of time over which a lead acid battery must be fully recharged determines the capacity of the battery more than the period of time over which it must supply power. Because of the need to reduce the charge rate – to avoid gassing – to a very low level as the battery approaches full charge, the last 15% of capacity takes a very long time to acquire (see Section 21.3).

19.11 Approval and certification

No base radio station may be operated in the UK prior to inspection of the installation by the Radio Investigation Service of the Radiocommunications Agency and the issue of their certificate of approval. The installation may not subsequently be modified without re-approval and certification.

References

Belcher R. *et al.* (1989). *Newnes Mobile Radio Servicing Handbook.* Butterworth-Heinemann, Oxford.

BS 6651: 1985. *Code of Practice for the Protection of Structures against Lightning.*

BS 6701: Part 1:1990. *Code of Practice for Installation of Apparatus Intended for Connection to Certain Telecommunications Systems.*

BS 7430: 1991. *Code of Practice for Earthing. Earthing of Telecommunications Installations.* International Telecommunications Union.

MPT 1331. *Code of Practice For Radio Site Engineering.*

MPT 1351. *Code of Practice For Repeater Operation at Communal Sites.*

MPT 1367. *Guidance Note to Legal Requirements Covering the Installation and use of Radio Apparatus and other Apparatus Generating Radio Frequency Emissions.*

MPT 1368. *Code of Practice for the Inspection of a Land Mobile Radio System for Conformity with the Wireless Telegraphy Act Licence and Performance Specification.*

Note: MPT documents are available from the Radiocommunications Agency (UK).

20 Instrumentation

20.1 Accuracy, resolution and stability

All measurements are subject to error and two instruments applied to the same piece of equipment under test may give a different answer. Tolerances must therefore be accepted. The errors arise from the following sources:

1. Human error, e.g. precision in reading a scale, use of incorrect instrument or range setting for the purpose.
2. The accuracy to which the instrument is able to display the result of a measurement or, in the case of a generator, the frequency or output level.
3. Accuracy of calibration.
4. Tolerances in the components used in the construction of the instrument. Variations in the load applied to the instrument.
5. Variations caused by long-term drift in the values of components.
6. Variations due to temperature and supply voltage fluctuations, and the warm-up time required by some instruments.
7. An effect on the circuit under test by the connection of the instrument.

There is an important difference between the accuracy and the resolution of an instrument. The accuracy is a statement of the maximum errors which may occur from the causes in statements 3 to 6 above. In instrument specifications, stability defined by statements 5 and 6 is usually quoted separately.

The accuracy of analogue measuring instruments is normally quoted as a percentage of full scale deflection (FSD). This is the accuracy of the instrument movement and components plus the scale calibration. The scale graduations, though, may not permit the user to determine the reading to the accuracy of the instrument perhaps because the graduations are cramped or parallax reading error occurs. These factors decide the resolution or precision of reading.

The accuracy of instruments with digital displays is usually quoted as a percentage of the reading plus or minus one count or one digit. While digital instruments are generally more accurate than their analogue counterparts, the fact that the least significant figure may be in error affects the resolution. *Figure 20.1* shows how this can arise. Most digital instruments use a gating process to switch the input to

Signal A

Signal B

Gate open

Gating pulse

Figure 20.1 Gating error introduced by signal phasing

the measuring circuitry for the appropriate period of time. The gating time itself may vary and affect the accuracy, but even with a perfectly accurate and stable gating time, the phase of the input signal at the time of switching affects the number of pulses passing through the gate and thus the resolution.

Performance figures taken from a number of manufacturer's catalogues are listed below. They show only the more important features of the specifications and are typical of those for high quality instruments used in radio work.

20.2 Audio instruments

20.2.1 Output power meters

Those for radio applications measure RMS audio power and are usually calibrated in watts and dBW or dBm. They usually contain a dummy load resistor of adjustable impedance as a substitute for the receiver loudspeaker when in use.

20.2.2 Distortion factor and Sinad meters

These may be separate instruments or combined, and may also be incorporated within an audio output power meter. Both operate on the same principle. An audio input amplifier, with automatic gain control to give a fixed reference level at its output, applies the output from the unit under test, first to a notch filter to remove the test frequency of 1000 Hz, then to an AC voltmeter. The voltmeter is calibrated in percentage distortion or dB Sinad.

In testing a receiver Sinad ratio, the instrument is connected across the loudspeaker or equivalent dummy load. An RF signal generator connected to the antenna socket is adjusted to apply a moderately high level signal modulated with a 1 kHz tone – which must be pure and match the notch filter in the Sinad meter – at 60% of the system peak modulation. To be correct, the receiver volume control is adjusted to give the rated audio output but if only Sinad sensitivity is being measured a lower level is acceptable. The Sinad meter will now display the combined level of all the frequencies except 1000 Hz present in the receiver's output. These are the noise and distortion products. With the high level input applied these will be low and the meter deflection will be small indicating a high Sinad ratio. The RF input level is now reduced until the meter reads 12 dB (the standard reference) and the RF level, μV or dBm, is the Sinad sensitivity for the receiver.

The oscillator to produce the test tone may be included in the meter and filters may be provided to comply with various weightings.

20.2.3 Audio signal generators

The instrument provides an output for synchronizing to an external standard and a TTL compatible output. A high power version capable of delivering 3 watts into a 3 Ω load in the audio band is available.

20.3 Radio frequency instruments

20.3.1 RF power meters

Direct reading RF power meters either contain a non-reactive load or use an external load and may be calibrated in watts or dBm. RF calorimeters convert the RF energy into heat and measure the temperature of the heated element. At low powers, 'dry' calorimeters are used but their long thermal time constant inhibits their use at high power levels. To measure high powers 'flow' calorimeters, where a fluid flows around a closed system and the output temperature of the fluid is measured, are used. Power can be determined from:

$$P = F(T_{out} - T_{in})c(T)$$

where

P = power
F = mass flow rate of the fluid
$c(T)$ = the fluid's specific heat
T_{in} = temperature of the fluid entering the load
T_{out} = temperature of the fluid after being heated by the load

Thruline type instruments require an external load which may be either an antenna system or a load resistor. This type of meter reads forward and reflected power and, in some instruments, VSWR directly.

20.3.2 RF signal generators

The range of instruments designed for use on specific systems, e.g. cellular and digitally modulated, is so wide that manufacturers' catalogues must be consulted for each application. In addition to the accuracy of the carrier frequency and output level, RF leakage, spectral purity and modulation noise levels are important. The output level may be calibrated in μV, dBV or dBm, and may refer to either an unterminated instrument (p.d.) or terminated in a load equal to the output impedance of the generator (e.m.f.). If the instrument is calibrated in p.d., the output voltage must be halved when the instrument is terminated in an equal impedance.

The 50 ohm output impedance of many signal generators matches the input impedance of most VHF and UHF receivers directly. To simulate the impedance of antennas at HF and below a dummy antenna is usually inserted between the generator and the receiver under test. *Figure 20.2* shows the circuit of a standard dummy antenna.

Figure 20.2   Standard dummy antenna

20.3.3 Frequency counters

Instruments are necessary to measure not only the precise frequencies of transmitters and receiver local oscillators but also those of low frequency signalling tones. Consequently they have a wide operating range.

The output of a transmitter must never be connected directly into a counter. The signal should either be obtained off-air or through a 'sniffer', appropriate for the frequency range, which siphons off a fragment of the transmitter output. It is also essential to ensure that the frequency displayed is the fundamental and not a harmonic or some other spurious frequency.

20.3.4 Modulation meters

The remarks in the above section relating to the connection of a transmitter and the selection of the fundamental frequency apply equally to modulation meters.

20.3.5 Spectrum analysers

Today's tight regulation of spectrum usage makes a spectrum analyser an essential tool for the precise alignment of radio equipment and for the measurement of modulation products and the noise content of signals.

A spectrum analyser is essentially a superheterodyne receiver with an adjustable IF bandwidth which sweeps across a portion of the frequency spectrum in synchronism with the horizontal trace of a cathode ray tube display. A signal at any frequency within the swept band will, while it remains within the IF passband, appear as a vertical displacement of the display trace. The design of the filter is crucial but must be a compromise. An analyser using multiple filters permits fast measurement but low resolution; a single filter offers high resolution but slower response time. Also, the shape factor of a wide filter inhibits the display of low level signals close to the centre frequency of the filter and a narrow filter, while permitting the display of these signals, may fail to display transients. The bandwidth determines the sweep speed, narrower bandwidths requiring slower speeds.

The possibility exists of spurious signals, e.g. intermodulation, being produced within the analyser with large input signals. A method of reducing the possibility is to insert a circulator and a notch filter

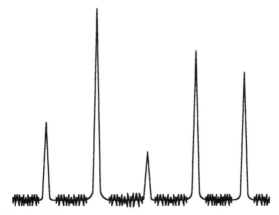

Figure 20.3 Amplitude-vs-frequency spectrum analyser display

between the unit under test and the analyser input to reduce the fundamental frequency amplitude. The circulator must be installed between the filter and the unit under test to absorb the reverse power produced by the impedance variations of the filter.

Many instruments are programmable and suitable for incorporation in automatic testing (ATE) systems, and instruments with similar features to those described are also offered in a combination test set form.

20.3.6 *Network analysers*

A network analysers examines *incident, reflected* and *transmitted* signals through a circuit or device, and displays the magnitude and phase of these signals. A spectrum analyser, on the other hand, measures only one channel, and displays magnitude and frequency.

Figure 20.4 shows an optical analogy to the network analyser. A 'device' of different optical density than ambient is in the path of an incident ray of light (R). When the light hits the surface, part of it is reflected (A) and part is transmitted through the 'devices' (even though refracted a bit).

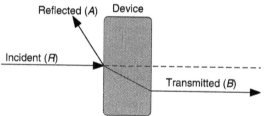

Figure 20.4 Optical analogy to network analyser

A network analyser consists of a three-channel RF receiver and a display. The incident signal is considered the *reference signal*, so is designated the R-channel. The other two channels receive the reflected signal on the A-channel, and the transmitted signal on the B-channel.

The uses of scalar and vector network analysers differ from the uses of spectrum analysers. The spectrum analyser measures external signals of unknown frequency and modulation type. Even when a tracking generator is added, to allow the spectrum analyser to perform stimulus-response tests, the spectrum analyser cannot do the job of the network analyser. The network analyser, by contrast, contains a known signal source, and is capable of sweeping a range of frequencies and power output levels. It can also perform ratio measurements.

References

Belcher, R. *et al.* (1989). *Newnes Mobile Radio Servicing Handbook*. Butter-worth-Heinemann, Oxford.
Bird Electronic Corp., Catalog GC-92.
Farnell Instruments Ltd, Catalogue, 1993.
Marconi Instruments Ltd, Test and Measurement Instrument Systems 1993/94.
RTT Systems Ltd, Catalogue, 1993.
Winder, S.W. (2001). *Newnes Telecommunications Pocket Book*. Butterworth-Heinemann, Oxford.

21 Batteries

21.1 Cell characteristics

Batteries are composed of cells which exhibit characteristics peculiar to their chemical constituents and construction. Common features, important to users, are their ability to store energy within a small space and with the least weight, and to release it at an adequate rate for the purpose under consideration.

21.1.1 Capacity

The amount of energy a battery can store is measured in ampere hours (Ah) at a specified discharge rate. For large cells this is usually the 10 hour rate but American practice, which is now almost universal, at least for smaller cells, is to use the 1 hour rate. It is important, therefore, to be certain which rate is referred to. The capacity reduces as the rate of discharge increases. Thus a battery of 60 ampere hours capacity at the 10 hour rate will provide 6 amps for ten hours before reaching the point at which it is considered to be discharged. If a current of 12 amps is taken, the battery will become discharged in less than 5 hours, and if the current is 3 amps, it will last longer than 20 hours. The rate of discharge is often referred to in terms of the C-rate which may be expressed in several ways. 1C, C or C_1 are numerically the same as the rated capacity, e.g. a 500 mA Nicad cell supplying 500 mA, which may be expressed as C, 1C or C_1 continuously will be discharged in approximately 1 hour. If the cell supplies current at a reduced rate, 0.5C, $0.5C_1$, or C/2, i.e. 250 mA, it will last approximately 125 minutes. The subscript, e.g. C_5, indicates the hourly discharge rate.

The terminal voltage at which a cell is considered discharged also varies with the discharge rate; a lead acid cell discharged at 1C is considered to be discharged when the terminal voltage falls to 1.75 V. At C_{10} the cell is considered to be discharged at 1.85 V.

The capacity of a battery or cell may also be specified at a given ambient temperature, usually 20°C. Lower temperatures reduce the effective capacity and maximum current off-take, higher temperatures increase them slightly.

For radio use battery duration may be quoted in terms of standby, receive and talk time. The duty cycle obviously varies from user to user but a useful standard for a radio-telephone is 90% standby, 5% receive and 5% transmit. On an open channel PMR system, 80%/15%/5% is more typical. Measuring the current drains during these activities enables the battery requirements to be calculated.

21.1.2 Internal resistance

The maximum instantaneous current which a battery can deliver is determined by its internal resistance. In this respect a battery behaves like any other generator (see Section 21.3.2), where increasing load currents produce an increasing voltage drop across the internal resistance. The internal resistance of a battery is seldom specified, but for a battery in good condition it is extremely low (one quoted figure is $15\,\mathrm{m}\Omega$ for a fully charged $500\,\mathrm{mAh}$ Nicad cell) and the voltage drop in the connecting leads will govern the maximum withdrawable current. Equally, the resistance of the meter used will affect the measurement of charge or discharge currents. If using an analogue meter, the older low resistance types are preferable (I keep a model 40 Avometer, $0.03\,\Omega$ int. res. on 12 amp range, for the purpose). Low current-rating fuses also present a resistance higher than that of a battery ($4\,\Omega$ measured for a $250\,\mathrm{mA}$ fuse, $0.5\,\Omega$ for a 1 A) and, probably, the connecting leads.

21.1.3 Power:weight and volume ratios

It is the battery that now limits the size to which radio equipment can be reduced. Recent developments have increased the power to weight and power to volume ratios which are possible. Typical ratios are referred to in the sections dealing with individual battery types.

21.1.4 Recharging conditions

The initial charge rate is the current flowing through a discharged battery to replace the charge in a specified time. Unless supplied from a constant current charger the current will fall as the battery voltage rises, but as full charge is approached the charge rate is usually reduced to a trickle or finishing charge rate.

Trickle charging maintains the cells in a fully charged condition by passing a very small current through them sufficient merely to replace any self-discharge losses through leakage.

Finishing charge is a rate to which the charging current is reduced when a battery reaches about 85% of its full capacity. It is a rate at which gassing is unlikely to occur.

Float charging maintains the cell voltage at its nominal while it is supplying continuous and variable loads.

21.2 Non-rechargeable, primary batteries

While rechargeable batteries are the obvious choice for use in equipment which is in use continuously, disposable types have economic and logistic advantages for some applications – the batteries are more expensive long term but no charger is needed. Also primary batteries have up to 4 times the available capacity of their nickel cadmium equivalents with less weight. Details of disposable batteries with some standard nickel-cadmium rechargeable equivalents are given in *Table 21.1*.

Lithium primary batteries are now available and have the following qualities:

1. A high cell voltage of 3.6 V on load. The voltage is constant – after an initial fall when load is first applied – until discharged when a rapid fall occurs.
2. An operating temperature range of $-55°C$ to $+75°C$ is possible.
3. Energy densities up to 630 mWh/g. A standard AA size cell has a power/weight ratio of 340 mWh/g.
4. Long shelf life. Ten years at room temperature for a 10% fall in capacity is envisaged.

They have a disadvantage in that fire, explosion or severe burns may result if the batteries are mistreated. They must not be recharged, disassembled, heated above 100°C, incinerated or the contents exposed to water.

One use of lithium batteries is as a power source for memories but protection against charging is needed. *Figure 21.1* shows a typical circuit.

Figure 21.1 Typical lithium battery application

Table 21.1 Cells and batteries

USA size	Nominal voltage (V)	Type†	IEC equivalent	Maximum dimensions (mm)			Contacts	Current (mA)	Weight (g)
				Length (or diameter)	Width	Height			
Zinc carbon									
N	1.5	D23	R1	12	–	00.1	Cap and base	1–5	7
AAA		HP 16	RO3	10.5	–	45	Cap and base	0–10 000	8.5
AA		HP7	R6	14.5	–	50.5	Cap and base	0–75	16.5
AA		C7	R6	14.5	–	50.5	Cap and base	0–75	16.5
C		SP11	R14	26.2	–	50	Cap and base	20–60	45
C		HP11	R14	26.2	–	50	Cap and base	0–1000	45
C		C11	R14	26.2	–	50	Cap and base	0–5	45
D		SP2	R20	34.2	–	61.8	Cap and base	25–100	90
D		HP2	R20	34.2	–	61.8	Cap and base	0–2000	90
	4.5	AD28	3R25	101.6	34.9	105	Socket	30–300	453.6
		1289	3R12	62	22	67	Flat springs	0–300	113
	6.0	PP8	4-F100-4	65.1	51.5	200.8	Press studs	20–151	1100
		PJ996	4-R25	67	67	102	Spiral springs	30–300	581
	9.0	991		135.7	72.2	125.4	Two screws	30–500	1470
		PP3-P	6-F22	26.5	17.5	48.5	Press studs	0–50	39
		PP3-C	6-F22	26.5	17.5	48.5	Press studs	0–50	39

(continued overleaf)

Table 21.1 (continued)

USA size	Nominal voltage (V)	Type†	IEC equivalent	Maximum dimensions (mm)			Contacts	Current (mA)	Weight (g)
				Length (or diameter)	Width	Height			
		PP3	6-F22	26.5	17.5	48.5	Press studs	0–10	38
		PP4	6-F20	26.5	–	50	Press studs	0–10	51
		PP6	6-F50-2	36	34.5	70	Press studs	2.5–15	142
		PP7	6-F90	46	45	61.9	Press studs	5–20	198
		PP9	6-F100	66	52	81	Press studs	5–50	425
		PP10	6-F100-3	66	52	225	Socket	15–150	1250
	15.0	B154	10-F15	16	15	35	End contacts	0.1–0.5	14.2
		B121	10-F20	27	16	37	End contacts	0.1–0.1	21
	22.5	B155	15-F15	16	15	51	End contacts	0.1–0.5	20
		B122	15-F20	27	16	51	End contacts	0.1–1.0	32
Manganese alkaline ED	1.5	MN1300*	LR20	34.2	–	61.5	Cap and base	10.00†	125
C		MN1400*	LR14	26.2	–	50	Cap and base	5.50†	65
AA		MN1500*	LR6	14.5	–	50.5	Cap and base	1.80†	23
AAA		MN2400*	LR03	10.5	–	44.5	Cap and base	0.80†	13
N		MN9100*	LR1	12	–	30.2	Cap and base	0.65†	9.6
Mercuric oxide	1.35/1.4	RM675H	NR07	11.6	–	5.4	Cap and base (button)	0.21†	2.6
		RM625N	MR9	15.6	–	6.2	Cap and base (button)	0.25†	4.3

Silver oxide		1.5	RM575H	NR08	11.6	—	3.5	Cap and base (button)	0.12†	1.4
			RM1H	NR50	16.4	—	16.8	Cap and base (button)	1.00†	12.0
			10L14	5R44	11.56	—	5.33	Cap and base (button)	0.13†	2.2
			10L124	5R43	11.56	—	4.19	Cap and base (button)	0.13†	1.7
			10L123	5R48	7.75	—	5.33	Cap and base (button)	0.08†	1.0
			10L125	5R41	7.75	—	3.58	Cap and base (button)	0.04†	0.8
Nickel cadmium		1.25	NC828	–		—		Button	0.28†	16.5
	AA		NCC50	–		See HP7		Button	0.60†	30.0
	C		NCC200	–		See HP11		Button	2.00†	78.8
	D		NCC400	–		See HP2		Button	4.00†	170.0
		10.0	NC828/8	–		—		Button stack	0.28†	125.0
		12.0	10/2250K	–		—		Button stack	0.225†	135.0
		9.0	TR7/8	(DEAC)		See PP3		Press studs	0.07†	45.0
	AA	1.25	501RS	(DEAC)		See HP7		Press studs	0.50†	30.0
	C		RS1.8	(DEAC)		See HP11		Press studs	1.80†	65.0
	D		RS4	(DEAC)		See HP2		Press studs	4.00†	150.0

† Capacity in ampere hours.
‡ BEREC types unless otherwise indicated.
* Also Duracell (Mallory).

21.3 Rechargeable batteries

21.3.1 Lead acid batteries

Lead acid batteries, whether of free electrolyte or low-maintenance sealed construction, have a nominal terminal voltage on load of 2.0 V per cell. This voltage falls on load in a gradual curve, shown in *Figure 21.2*, until the discharged voltage of between 1.75 and 1.85 volts per cell is reached.

Figure 21.2 Typical lead acid cell discharge characteristics

A discharged battery will, because of its inefficiencies, require a recharge equal to the amperes × hours discharged +11%, e.g. a cell discharged at 5 amps for 10 hours will require a recharge of 55.5 ampere hours with a constant current at the 10 hour rate. Because of the reducing current as full charge is approached a recharge time of 1.4 to 1.5 times the capacity to be restored is more practical. The final on-charge cell voltage can increase to approximately 2.7 volts. Gassing occurs and hydrogen is liberated when the cell voltage reaches 23 V but provided the charging current is sufficiently low above this point gassing will be avoided. This lower charge rate, the 'finishing rate' can be applied by maintaining the charging voltage at about 2.4 volts when the battery will automatically limit the charging current. The specific gravity of the electrolyte in a fully charged cell is between 1.205 and 1.215.

The trickle charge current must be low enough to avoid gassing. A current of 7% of the 10 hour capacity is typical.

Float charging should maintain a cell voltage of approximately 2.2 volts.

The power:weight ratio of lead acid batteries is poor, a small (4 Ah), 6.0 V, sealed lead acid battery having a power:weight ratio of 26 mWh/g.

21.3.2 Nickel-cadmium (Nicad) batteries

At present the Nicad is probably the most commonly used rechargeable battery for portable applications. A standard size AA cell has a power:weight ratio of 27 mWh/g. The on-load Nicad cell voltage, after an initial fall from the on-charge voltage of between 1.3 and 1.4 V, remains substantially constant at about 1.2 V until the discharged voltage of 0.9 to 1.1 V is reached. Thereafter the voltage falls rapidly. This is illustrated in *Figure 21.3*. While the constant voltage is ideal during discharge it poses a problem in that it is difficult to reliably measure the intermediate state of charge which created difficulties with recharging. A Nicad battery which is repeatedly partially discharged and then recharged may, after many cycles, behave as though it were fully discharged when the repeated recharge condition is reached (the memory effect) and, with a fixed time charger, the possibility of over-charging is present. One solution fully discharged all batteries after use to a predetermined level, typically 1.1 V per cell, and then recharged them at a constant current for a fixed period of time. Unfortunately, this procedure shortened the life of the batteries; *Figure 21.4* shows the life expectancy of a cell with repeated discharges. The present solution is to charge the batteries automatically to the fully charged state and then reduce the current to the trickle charge level. Batteries which are subjected to repeated partial discharge may then be occasionally fully discharged to obviate the memory effect.

Figure 21.3 Typical Nicad cell discharge characteristics

Constant current, automatic charging is recommended. Chargers vary in complexity, some detecting the end-of-charge point by sensing a variation of voltage. At end of charge the cell voltage first rises and then falls slightly as in *Figure 21.5*. More sophisticated chargers also sense the cell case temperature which rises during charge. Batteries are available for standard charging at the ten hour rate where 14 to

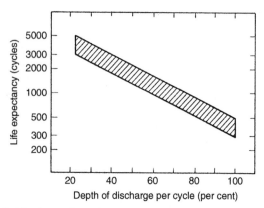

Figure 21.4 Nicad cell: effect of repeated discharge vs. cell life

(a) Standard charge, C/10 × 15 hrs

(b) Rapid charge, 1 C × 90 mins

Figure 21.5 Nicad cell: discharging characteristics

15 hours will be required to recharge a fully discharged battery, fast charging at 5 C and rapid charging at C rate.

Cells may be fitted with a re-sealing one-way vent which opens at about 200 psi and closes at about 175 psi to relieve any excess internal pressure caused by a fault or abuse.

22 Satellite communications

22.1 Earth orbits

Communications satellites are required to illuminate the earth with radio signals and their orbits are chosen according to the size and location of the part of the earth's surface they must light up.

A satellite orbiting the earth is continuously pulled by a centripetal force, in this case gravity, towards the centre of the earth. It is also pulled by centrifugal force to leave its orbit at a tangent. When these opposing forces are equal in magnitude the satellite is in a stable orbit. There is, then, for a given height (the radius of the path minus the radius of the earth, 6378 km), a velocity at which the conditions for stable orbit apply, and which determines the orbiting time.

22.1.1 Geostationary orbits

Satellites relay information from ground stations, either fixed or mobile, or between satellites. It is an advantage for some purposes, therefore, to use satellites with an orbit time identical to that of the earth so that no tracking from the ground stations is needed. Geostationary satellites have the same angular velocity as the earth making them appear to be stationary. Their height is 35 788 km and four such satellites cover the earth from latitude 81.3°N to 81.3°S as in *Figure 22.1*.

The disadvantages of geostationary satellites are that they are in a high earth orbit (HEO) resulting in a signal delay of 240 ms for the complete go and return path. Also they are in an equatorial orbit so that signals to the higher latitudes travel at a shallow angle to the earth's surface rendering them unsuitable for mobile use where communications must be achieved at street level in cities.

As each satellite covers a large portion of the earth, the design of their antennas to permit repeated frequency re-use is important and antennas with small, steerable footprints have been developed for this purpose. An advantage, in addition to the lack of tracking, is that the shadowing by the earth is minimal so that solar power cells receive almost continuous illumination.

22.1.2 Elliptical orbits

A polar orbit, where the satellite follows a North/South track, provides the opportunity to survey the earth in a series of strips. The satellites

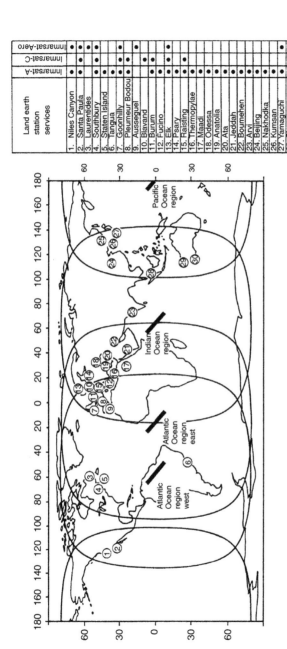

Land earth station services	Immarsat-A	Immarsat-C	Immarsat-Aero
1. Niles Canyon	•		•
2. Santa Paula	•	•	•
3. Laurentides			•
4. Southbury	•	•	•
5. Staten Island	•		
6. Tangua	•		
7. Goonhilly	•	•	•
8. Pleumeur Bodou	•	•	
9. Aussequel			•
10. Blavand	•	•	
11. Burum	•	•	
12. Fucino	•		•
13. Elk	•		
14. Psary	•		
15. Raisting	•	•	
16. Thermopylae	•		
17. Maadi	•		
18. Odessa	•		
19. Anatolia	•		
20. Ata	•		
21. Jeddah	•		
22. Boumehen	•		
23. Arvi	•		
24. Beijing	•		
25. Nakhodka	•		
26. Kumsan	•		
27. Yamaguchi	•		•
28. Singapore	•	•	•
29. Perth	•	•	•
30. Ganagara	•		

Figure 22.1 Immarsat coverage and GES locations

used for this purpose are in low each orbit (LEO) and consequently have a high velocity. Their orbit time is approximately 1.5 hours and between successive orbits the earth has rotated 22.5°. Sixteen orbits are therefore needed to scan the earth's surface. Until recently polar orbits were used only for optical surveillance but now several projects for radio communication are either in the early stages of installation or under development. These use a number of satellites so that for mobile communications, for instance, there is a satellite continuously in view. Tracking of these extremely fast satellites by the ground stations, which may themselves be moving, along with Doppler effect has been a major obstacle which is now being overcome.

22.2 Communications by satellite link

Satellites are radio links and receive signals from the ground and other satellites which they must re-transmit. The signals from ground stations are comparatively weak and require high power amplification for onward transmission. As the satellite's receive and transmit antennas must be close together, the possibility of RF instability prohibits on-frequency repetition. The up-link, from ground station to satellite, must therefore be converted to another before re-transmission. The up-link frequency is normally higher than the down-link and the frequency converter is referred to as a 'down-converter'.

The frequency bands used for communications purposes are listed in *Table 22.1*.

Satellite television

The broadcasting of television may be via either communications satellites or Direct Broadcasting by Satellite (DBS) satellites. The positions of non-DBS satellites relative to the UK are shown in *Figure 22.2* and *Table 22.2* lists the European channels. *Figure 22.3* shows the world allocations of DBS satellites and *Tables 22.3* and *22.4* list the channel frequencies and national allocations. The frequency plan for the Astra satellite is in *Table 22.5*.

22.3 Proposed satellite television formats

Most current European satellite television programmes (non-DBS) are broadcast as fairly standard PAL signals, FM modulated into the satellite channel. DBS transmissions, and those from Astra satellite, will probably be of a multiplexed analogue component (MAC) format. In MAC, data corresponding to sound tracks and subtitles, etc., an

Table 22.1 Communications satellite frequencies

Frequency band (GHz)	Link	European telecom links	International telecom links
1.5–1.6	Down	Mobile	
1.6–1.7	Up	Mobile	
3.4–4.2	Down	Fixed	Fixed
4.5–4.8	Down	Fixed	Fixed
5.9–7.0	Up	Fixed	Fixed
10.7–11.7	Down	Fixed (+ Non-DBS television)	
11.7–12.5	Down	Fixed (+ television)	
12.5–12.75	Down	Fixed (private links)	
12.75–13.75	Up	Fixed (private links)	
14.0–14.8	Up	Fixed	
17.3–18.3	Up	Fixed	
17.7–20.2	Down	Fixed	
20.2–21.2	Down	Mobile	
27.0–30.0	Up	Fixed	
30.0–31.0	Up	Mobile	
22.5–23.55		Allocated for intersatellite links	
32.0–33.0		Allocated for intersatellite links	
54.25–58.2		Allocated for intersatellite links	
59.0–64.0		Allocated for intersatellite links	

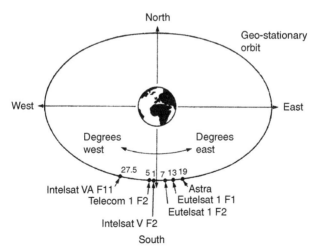

Figure 22.2 Non-DBS satellite television: positions of the main non-DBS satellites relevant to the UK

Table 22.2 European satellite television channels broadcast via communications satellites

Channel	Frequency (GHz)	Polarization	Audio sub-carrier	Video standard	Satellite
3Sat	11.175	H	6.65	PAL	Eutelsat 1 F1
Anglovision	11.515	V	6.6	PAL	Intelsat VA-F11
Arts Channel	11.135	H	6.6	PAL	Intelsat VA-F11
BBC1/2	11.175	H	6.65	PAL	Intelsat VA-F11
CanalJ	12.564	V	5.8	PAL	Telecom 1 F2
Children's Channel	11.015	H	6.6	PAL	Intelsat VA-F11
CNN	11.155	V	6.65	PAL	Intelsat VA-F11
Filmnet	11.140	V	6.6	PAL	Eutelsat 1 F1
Moscow 1	3.675	–	7.0	SECAM	Gorizont-12
Moscow 2	3.0	–	7.0	SECAM	Gorizont-7
Infofilm & Video	11.015	H	6.6	PAL	Intelsat F2
La Cinq	12.606	V	5.8	SECAM	Telecom 1 F2
Lifestyle	11.135	H	6.6	PAL	Intelsat VA-F11
M6	12.648	V	5.8	SECAM	Telecom 1 F2
MTV	10.975	H	6.65	PAL	Intelsat VA-F11
Norsk Rikskringkasti	11.644	H	Digital	C-MAC	Eutelsat 1 F2
Premiere	11.015	H	6.6	PAL	Intelsat VA-F11
RAI-Uno	11.007	H	6.6	PAL	Eutelsat 1 F1
RTL-Plus	11.091	V	6.65	PAL	Eutelsat 1 F1
SAT1	11.507	V	6.65	PAL	Eutelsat 1 F1
Satellite Information	11.575	H	Digital	B-MAC	Intelsat VA-F11
Screensport	11.135	H	6.6	PAL	Intelsat VA-F11
Skychannel	11.650	H	6.65	PAL	Eutelsat 1 F1
Superchannel	10.674	V	6.65	PAL	Eutelsat 1 F1
SVT-2	11.178	H	Digital	C-MAC	Intelsat F2
SVT-2	11.133	H	Digital	C-MAC	Intelsat F2
Teleclub	11.987	V	6.5	PAL	Eutelsat 1 F1
TV5	11.472	H	6.65	PAL	Eutelsat 1 F1
Worldnet	11.512	H	6.65	PAL	Eutelsat 1 F1
Worldnet	11.591	H	6.6	SECAM	Eutelsat 1 F1
Worldnet	12.732	V	5.8	NTSC	Telecom 1 F2

analogue signal corresponding to chrominance and an analogue signal corresponding to luminance are transmitted separately in each broadcast line of the picture.

In order to achieve multiplexing of the three parts of the format, time compression of chrominance and luminance signals occurs before

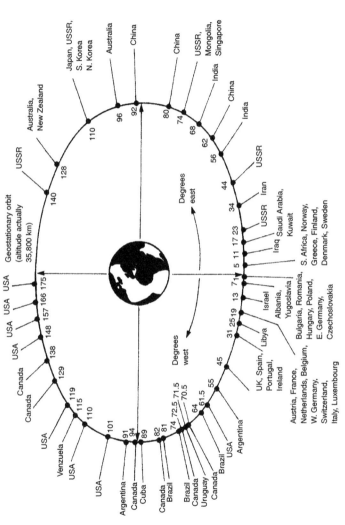

Figure 22.3 World allocations of DBS satellite positions

Table 22.3 DBS television channels

Channel number	Frequency (GHz)		
1	11.72748	21	12.11108
2	11.74666	22	12.13026
3	11.76584	23	12.14944
4	11.78502	24	12.16862
5	11.80420	25	12.18780
6	11.82338	26	12.20698
7	11.84256	27	12.22616
8	11.86174	28	12.24534
9	11.88092	29	12.26452
10	11.90010	30	12.28370
11	11.91928	31	12.30288
12	11.93846	32	12.32206
13	11.95764	33	12.34124
14	11.97682	34	12.36042
15	11.99600	35	12.37960
16	12.01518	36	12.39878
17	12.03436	37	12.41796
18	12.05354	38	12.43714
19	12.07272	39	12.45632
20	12.09190	40	12.47550

transmission, and they must be reconstituted at the receiver. Two main variations of the MAC format have been selected for European broadcasters, D-MAC and D2-MAC. They differ basically in the number of data channels, and hence the overall bandwidth required. Both modulate video signals in FM, and data signals in duobinary FM.

22.4 Global positioning system (GPS)

This system uses an American government satellite network comprised of 24 satellites in three circular orbits at a height of 20 000 km and with twelve-hour periods. At least six satellites should be visible from any point around the earth at any one time.

Table 22.4 Direct broadcast by satellite (DBS) television. European allocation of satellite positions, channel and polarizations

Band	5° E	Orbital position 19° W	31° W	37° W
11.7–12.1 GHz RH polarized	*Turkey:* CH. 1, 5, 9, 13, 17 *Greece:* CH. 3, 7, 11, 15, 19	*France:* CH. 1, 5, 9, 13, 17 *Luxembourg:* CH. 3, 7, 11, 15, 19	*Eire:* CH. 2, 6, 10, 14, 18 *UK:* CH. 4, 8, 12, 16, 20	*San Marino:* CH. 1, 5, 9, 13, 17 *Lichtenstein:* CH. 3, 7, 11, 15, 19
11.7–12.1 GHz LH polarized	*Finland:* CH. 2, 6, 10 *Norway:* CH. 14, 18 *Sweden* CH. 4, 8 *Denmark:* CH. 12, 16, 20	*Germany:* CH. 2, 6, 10, 14, 18 *Austria:* CH. 4, 8, 12, 16, 20	*Portugal:* CH. 3, 7, 11, 15, 19	*Andorra:* CH. 3, 8, 12, 16, 20
12.1–12.5 GHz RH polarized	*Cyprus:* CH. 21, 25, 29, 33, 37 *Iceland, etc.:* CH. 23, 27, 31, 35, 39	*Belgium:* CH. 21, 25, 29, 33, 37 *Netherlands:* CH. 23, 27, 31, 35, 39		*Monaco:* CH. 21, 25, 29, 33, 37 *Vatican* CH. 23, 27, 31, 35, 39
12.1–12.5 GHz LH polarized	*Nordic group** CH. 22, 24, 26, 28 30, 32, 36, 40 *Sweden:* CH. 34 *Norway:* CH. 38	*Switzerland:* CH. 22, 26, 30, 34, 38 *Italy:* CH. 24, 28, 32, 36, 40	*Iceland:* CH. 21, 25, 29, 33, 37 *Spain:* CH. 23, 27, 31, 35, 39	

*Wide beam channels: Denmark, Finland, Norway, Sweden.

Table 22.5 Astra television channels

Channel	Astra 1-A (GHz)	Channel	Astra 1-B (GHz)	Channel	Astra 1-C (GHz)	Channel	Astra 1-D (GHz)
1	11.21425	17	11.46425	33	10.96425	49	10.71425
2	11.22900	18	11.47900	34	10.97900	50	10.72900
3	11.24375	19	11.49375	35	10.99375	51	10.74375
4	11.25850	20	11.50850	36	10.00850	52	10.75850
5	11.27325	21	11.52325	37	10.02325	53	10.77325
6	11.28800	22	11.53800	38	10.03800	54	10.78800
7	11.30275	23	11.55275	39	10.05275	55	10.80275
8	11.31750	24	11.56750	40	10.06750	56	10.81750
9	11.33225	25	11.58225	41	10.08225	57	10.83225
10	11.34700	26	11.59700	42	10.09700	58	10.84700
11	11.36175	27	11.61175	43	10.11175	59	10.86175
12	11.37650	28	11.62650	44	10.12650	60	10.87650
13	11.39125	29	11.64125	45	10.14125	61	10.89125
14	11.40600	30	11.65600	46	10.15600	62	10.90600
15	11.42075	31	11.67075	47	10.17075	63	10.92075
16	11.43550	32	11.68550	48	10.18550	64	10.83550

Each satellite transmits continuously updated information about its orbit on two frequencies, 1227 MHz and 1575 MHz. One radio channel carries two pseudo-random codes, one very long and the other very short. The second channel is modulated only with the short code. The codes enable the satellite to be positively identified, and the distance from a receiver on, or close to, the earth to be calculated.

A receiver's position, in three-dimensional space, is identified by measuring and calculating the distance from three satellites. The short codes provide the initial fix but increased precision is obtained from the long codes. There is an error due to time variations arising from various sources of which satellite speed is one, but taking a measurement from a fourth satellite enables a correction factor to be applied.

Receivers which utilize only the short PN codes provide a resolution accurate to about 100 m. Those which can process the long codes provide a fix accurate to about 45 m.

References

Lewis, G.E. (1988). *Communication Services via Satellite*. BSP Professional Books, Oxford.

Long, M. *The 1993/1994 World Satellite Annual*. Mark Long Enterprises Inc., Ft Lauderdale, USA.

23 Connectors and interfaces

23.1 Audio and video connectors

Audio connectors

The DIN standards devised by the German Industrial Standards Board are widely used for the connection of audio equipment. The connectors are shown below. The 3-way and 5-way 45 are the most common and connections for those are listed.

	Mono	Stereo
Microphone	Input 1	Input LH 1
	O V 2	Input RH 4
		O V 2
	Pin 3 available	Pins 3 and 5
	for polarizing	available for
	voltage	polarizing voltage
Tape recorder	Input 1	Input LH 1
inputs and	O V 2	Input R4 4
monitor outputs	Output 3	O V 2
		Output LH 3
		Output RH 5

	Mono	*Stereo*
Tape recorder replay output	Output, low Z1	Output LH, low Z1
	O V 2	Output RH, low Z4
	Output, high Z3	O V 2
		Output LH, high Z3
		Output RH, high Z5
Amplifiers	Output to tape 1	Output LH 1
	O V 2	Output RH 1
	Input from tape 3	O V 2
		Input LH 3
		Input RH 5

Variations on the above exist between different manufacturers.

Videorecorder/televisions/camera connectors

Standard pin configurations for videorecorders, televisions and videocameras are shown below. Many follow standard DIN connector pinouts, but videocamera and SCART connectors differ significantly.

(a) 5-pin DIN   (b) 6-pin DIN   (c) 7-pin DIN

1,4 Audio in
2   Chassis
3,5 Audio out
    or stereo
1   L audio in
2   Chassis
3   L audio out
4   R audio in
5   R audio out

1 AV select,
  For VCR
  high = all
  outputs,
  low = all
  inputs,
  Opposite
  for TV set
2 Video in/out
3 Chassis
4 L audio in/
  out
5 12 V
6 R audio in/
  out

1 L audio in
2 Chassis
3 L audio out
4 R audio in
5 R audio out
6 Remote
  control
  data
7 Chassis

SCART (BS 6552)

SCART connectors, also known as Peritelevision or Euroconnector connectors, feature two control systems which allow remote control over the television's or videorecorder's functions.

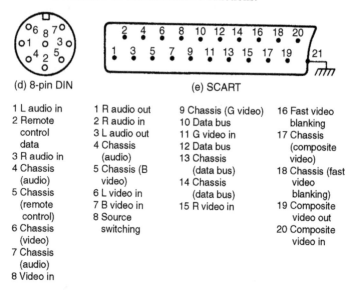

(d) 8-pin DIN   (e) SCART

1 L audio in
2 Remote control data
3 R audio in
4 Chassis (audio)
5 Chassis (remote control)
6 Chassis (video)
7 Chassis (audio)
8 Video in

1 R audio out
2 R audio in
3 L audio out
4 Chassis (audio)
5 Chassis (B video)
6 L video in
7 B video in
8 Source switching

9 Chassis (G video)
10 Data bus
11 G video in
12 Data bus
13 Chassis (data bus)
14 Chassis (data bus)
15 R video in

16 Fast video blanking
17 Chassis (composite video)
18 Chassis (fast video blanking)
19 Composite video out
20 Composite video in

The simplest is a source switching input (pin 8) in which an external source (videorecorder, computer, etc.) can, by issuing a 12 volt signal, cause the television to switch to baseband inputs.

A more complex control system, called *domestic data bus* (D^2B), is given through pins 10 and 12, in which serial data can be passed between controlling microprocessors in the television and external equipment. No standard yet exists for D^2B.

23.2 Co-axial connector

The most commonly used connectors for RF cables are:

Type	Impedance, Z_0 (ohms)	Max. VSWR to (frequency GHz)	Maximum proof RF voltage	Notes
N	50 or 75	1.30 (4)	1.5 kV (5 MHz at sea level)	Screw together

Type	Impedance, Z_0 (ohms)	Max. VSWR to (frequency GHz)	Maximum proof RF voltage	Notes
BNC	50 or 75	1.01 (1) 1.30 (4)	1 kV (5 MHz at sea level)	Miniature. Bayonet fitting
TNC	50 or 75	As BNC	As BNC	Miniature. Robust screw together
SMB	50	1.46 (4)	1.5 kV (At sea level. Cable dependent)	Sub-miniature. Snap together
SMC	50	1.41 (4) 1.69 (10)	As SMB	Sub-miniature. Screw together
SMD	50	As SMB	As SMB	Sub-miniature. Push together
BT 43	75			Developed from SMB range for use in telecomms and data transmission
7–16	50	1.3 (5 GHz)	2.7 (connector)	Suitable for medium to high power applications in the cellular and broadcast industries. Screw together
PL259/ SO239/ UHF 83	50	–	500 (pk)	Non-constant imp. High VSWR makes it unsuitable for use above 144 MHz and for extending RF cables. Very robust. Screw together
C	75	–	–	Bayonet fitting

Type	Impedance, Z_0 (ohms)	Max. VSWR to (frequency GHz)	Maximum proof RF voltage	Notes
F	50	–	–	American CCTV connector used on some 144 MHz hand-portable transceivers. Plugs use inner conductor of cable as centre pin
Belling Lee	50	–	–	British TV antenna connector. Aluminium versions may corrode when used outdoors
GR	50	1 MHz max. frequency	–	Constant imp. sexless connector
Phono	–	–	–	American connector for audio use

Examples of assembly instructions for co-axial RF connectors (by kind permission of M/ACOM Greenpar Ltd.)

 Measuring instrument – a rule is shown, but better results are obtained by using a Vernier gauge

 Stout trimming blade, suitable for cutting copper wire braid

 Crimping tool

Soldering iron

Side cutters, also for trimming braid

Spanner, of the relevant size for the connector

Small screwdriver

Hacksaw, sometimes appropriate for semi-rigid cable, although for repetitive operations a power trimmer should be considered

Assembly instructions Type N

Cable types:
  50 ohm: PSF1/4M (BBC), RG 8A/U, RG 213/U, URM 67
  75 ohm: RG 11A/U, RG 63B/U, RG 114A/U, URM 64

1 Side clamp nut and plain gasket over cable and trim outer sheath from cable as shown.

Clamp nut   Plain gasket

2 Fold back braid and push ferrule over dielectric to trap braid between outer sheath and ferrule. Trim off surplus braid.

Ferrule

262

3 Trim back dielectric and check the length of the protruding centre conductor.

4 Tin centre conductor, then slide rear insulator over dielectric, to butt against ferrule.

Real insulator

5 Fit contact (male for plugs, female for jacks) onto centre conductor. Hold cable and contact tightly together and solder.

Male contact

6 Slide plain gasket and clamp nut up to ferrule trapping braid. Fit front insulator over contact to butt against rear insulator and press sub-assembly into body as far as possible.

Front insulator   Body

7 Engage and tighten clamp nut.

1 Slide crimp sleeve over cable

Crimp sleeve

2 Trim back outer sheath and braid to dimensions shown. Any foils adjacent to the dielectric should be left in place but trimmed 1.5 mm from the face of the dielectric

18.2
4.7
10.3

3 Fit contact (male for plugs, female for jacks) over centre conductor to butt against dielectric and crimp.

Male contact

4 Push fit sub-assembly into body until contact clicks into insulator. In all cases, braids (and any intermediate foils) slide over ferrule while any foil adjacent to the dielectric should enter the ferrule bore.

Body

5 Slide crimp sleeve forward over braid and ferrule to butt against rear of the connector and crimp.

Assembly instructions Type BNC

Cable types:
RG 58C/U, RG 141A/U, URM 43, URM 76

1 Slide clamp nut and plain gasket over cable and trim outer sheath from cable, as shown.

Clamp nut   Plain gasket

2 Fold back braid and push ferrule over dielectric to trap braid between outer sheath and ferrule. Trim off surplus braid.

Ferrule

3 Trim back dielectric and check the length of the protruding centre conductor.

4 Tin centre conductor, then slide rear insulator over dielectric, to butt against ferrule.

Rear insulator

5 Fit contact (male for plugs, female for jacks) onto centre conductor, with collar pressed into recess in rear insulator. Hold cable and contact tightly together, and solder.

Male contact

6 Slide plain gasket and clamp nut up to ferrule, trapping braid. Fit front insulator over contact to butt against rear insulator and press sub-assembly into body as far as possible.

Front insulator

Body

7 Engage and tighten clamp nut

1 Slide metal crimp sleeve over cable, trim outer sheath from cable as shown.

16

Sleeve

2 Trim back braid and dielectric to the dimensions shown.

8.0

4.0

3 Fit contact over centre conductor to butt against the dielectric, then crimp.

Male contact

4 Press sub-assembly into body, until contact clicks into place and ensuring that the knurled ferrule is inserted between the dielectric and braid.

5 Slide the sleeve along the cable, until it butts against the body sub-assembly. Crimp, using the tool listed below.

Note: A plug is shown, but these instructions are relevant to both plugs and jacks. The shape of contacts and insulators may also vary from the drawings shown.

Assembly instructions Types SMB/SMC/SMD

Cable types:

TM 3306, RG174A/U, RG188A/U, RG316/U
TM 3263, RG178B/U, RG196A/U, URM 110

1 Slide clamp nut, a washer, a gasket and the other washer over the cable trim outer sheath to dimension shown.

2 Fold back braid. Push ferrule over dielectric to trap braid between outer sheath and ferrule.

Ferrule

3 Trim off surplus braid. Trim dielectric flush with ferrule and check length of centre conductor. Tin centre conductor.

1.5

Real insulator

4 Side rear insulator over centre conductor until it butts against ferrule.

Contact

5 Fit contact onto centre conductor until it butts against rear insulator. Hold cable and contact tightly together and solder.

Front insulator

6 Fit front insulator over contact until it butts against internal shoulder.

Body

7 Press sub-assembly into body as far as possible. Engage and tighten clamp nut.

1 Place crimp sleeve over sheat and trim cable to the dimensions shown. Ensure that the centre conductor is not damaged.

Crimp sleeve

2 Place contact on centre conductor and crimp.

Contact

3 Push contact into body sub-assembly. Ensure the contact 'clicks' into rear insulator with the ferrule between the braid and the dielectric.

4 Slide crimp sleeve over the braid and crimp.

23.3 Interfaces

23.3.1 Connectors and connections

Data interchange by modems

When transmitting and receiving data across telephone or other circuits, the equipment which actually generates and uses the data (e.g. a computer or VDU terminal) is known as *data terminating equipment* (DTE). The equipment which terminates the telephone line and converts the basic data signals into signals which can be transmitted is known as *data circuit-terminating equipment* (DCE). As far as the user is concerned the interface between DTE and DCE is the most important. ITU-T recommendation V24 defines the signal interchanges and functions between DTE and DCE; these are commonly known as the 100 series interchanges circuits:

Interchange circuit		Data		Control		Timing	
Number	*Name*	*From DCE*	*To DCE*	*From DCE*	*To DCE*	*From DCE*	*To DCE*
101	Protective ground or earth						
102	Signal ground or common return						
103	Transmitted data		●				
104	Received data	●					
105	Request to send				●		
106	Ready for sending			●			
107	Data set ready			●			
108/1	Connect data set to line				●		
108/2	Data terminal ready				●		
109	Data channel received line signal detector			●			
110	Signal quality detector			●			
111	Data signalling rate selector (DTE)				●		
112	Data signalling rate selector (DCE)			●			
113	Transmitter signal element timing (DTE)						●

| Interchange circuit | | Data | | Control | | Timing | |
Number	Name	From DCE	To DCE	From DCE	To DCE	From DCE	To DCE
114	Transmitter signal element timing (DCE)					•	
115	Receiver signal element timing (DCE)					•	
116	Select stand by				•		
117	Standby indicator			•			
118	Transmitted backward channel data		•				
119	Received backward channel data	•					
120	Transmit backward channel line signal				•		
121	Backward channel ready			•			
122	Backward channel received line signal detector			•			
123	Backward channel single quality detector			•			
124	Select frequency groups				•		
125	Calling indicator			•			
126	Select transmit frequency				•		
127	Select receive frequency				•		

Number	Name	Data From DCE	Data To DCE	Control From DCE	Control To DCE	Timing From DCE	Timing To DCE
128	Receiver signal element timing (DTE)						•
129	Request to receive			•			
130	Transmit backward tone			•			
131	Received character timing					•	
132	Return to non-data mode			•			
133	Ready for receiving			•			
134	Received data present	•					
191	Transmitted voice answer			•			
192	Received voice answer	•					

Modem connector pin numbers

The connectors used with 100 series interchange circuits and its pin assignments are defined by international standard ISO 2110 and are (for modems following the ITU-T recommendations V21, V23, V26, V26bis, V27 and V27bis) as follows:

Pin number	Interchange circuit numbers			Pin Name
	V21	V23	V26/V27	
1	*1	*1	*1	Ground
2	103	103	103	TXD
3	104	104	104	RXD
4	105	105	105	RTS
5	106	106	106	RFS
6	107	107	107	DSR
7	102	102	102	Signal return
8	109	109	109	Signal DET
9	*N	*N	*N	
10	*N	*N	*N	
11	126	*N	*N	STF
12	*F	122	122	
13	*F	121	121	
14	*F	118	118	
15	*F	*2	114	
16	*F	119	119	
17	*F	*2	115	
18	141	141	141	
19	*F	120	120	
20	108/1-2	108/1-2	108/1-2	DTR
21	140	140	140	
22	125	125	125	Call ind.
23	*N	111	111	
24	*N	*N	113	
25	142	142	142	

Notes:

*1 Pin 1 is assigned for connecting the shields between tandem sections of shielded cables. It may be connected to protective ground or signal ground.

*F Reserved for future use.

*N Reserved for national use.

9-Pin connector

Pin number	Pin name	Description
1	DCD	Data carrier detect
2	RXD	Receive data
3	TXD	Transmit data
4	DTR	Data terminal ready
5	GND	Ground
6	DSR	Data set ready
7	RTS	Ready to send
8	CTS	Clear to send
9	RI	Ring indicator

Automatic calling

A similar series of interchange circuits is defined in ITU-T recommendation V25 for automatic calling answering between modems over the telephone network. This is the 200 series interchange circuits:

Interchange circuit

Number	Name	From DCE	To DCE
201	Signal ground	•	•
202	Call request		•
203	Data line occupied	•	
204	Distant station connected	•	
205	Abandon call	•	
206	Digit signal (2^0)		•
207	Digit signal (2^1)		•
208	Digit signal (2^2)		•
209	Digit signal (2^3)		•
210	Present next digit	•	
211	Digit present		•
213	Power indication	•	

RS 232C

The EIA equivalent of CCITT V24 interface is the RS 232C specification, which similarly defines the electrical interface between DTE and DCE. Although the two have different designations, they are to all practical purposes equivalent. The RS 232C interchange circuits are:

Mnemonic	Name	Data		Control		Timing	
		From DCE	To DCE	From DCE	To DCE	From DCE	To DCE
AA	Protective ground						
AB	Signal ground/common return						
BA	Transmitted data		•				
BB	Received data	•					
CA	Request to send				•		
CB	Clear to send			•			
CC	Data set ready			•			
CD	Data terminal ready				•		
CE	Ring indicator			•			
CF	Received line signal detector			•			
CG	Signal quality detector			•			
CH	Data signal rate selector (DTE)				•		
CI	Data signal rate selector (DCE)			•			
DA	Transmitter signal element timing (DTE)						•
DB	Transmitter signal element timing (DCE)					•	
DD	Receiver signal element timing (DCE)					•	

Mnemonic	Name	Data		Control		Timing	
		From DCE	To DCE	From DCE	To DCE	From DCE	To DCE
SBA	Secondary transmitted data		●				
SBB	Secondary received data	●					
SCA	Secondary request to send				●		
SCB	Secondary clear to send			●			
SCF	Secondary received line signal detector			●			

RS 449

The EIA RS 232C standard, although the most common, is by no means perfect. One of its main limitations is the maximum data rate −18.2 K baud. Various improved interchange circuits (RS 422, RS 423) have been developed. The RS 449 standard is capable of very fast data rates (up to 2 Mbaud):

Mnemonic	Name	Data		Control		Timing	
		From DCE	To DCE	From DCE	To DCE	From DCE	To DCE
SG	Single ground						
SC	Send common						
RC	Receive common						
IS	Terminal in service				●		
IC	Incoming call			●			
TR	Terminal ready				●		

Interchange circuit		Data		Control		Timing	
Mnemonic	Name	From DCE	To DCE	From DCE	To DCE	From DCE	To DCE
DM	Data mode			•			
SD	Send data		•				
RD	Receive data	•					
TT	Terminal timing						•
ST	Send timing					•	
RT	Receive timing					•	
RS	Request to send				•		
CS	Clear to send			•			
RR	Receiver ready			•			
SQ	Signal quality			•			
NS	News signal				•		
SF	Select frequency				•		
SR	Signalling rate selector				•		
SI	Signalling rate indicator			•			
SSD	Secondary send data		•				
SRD	Secondary receive data	•					
SRS	Secondary request to send				•		
SCS	Secondary clear to send			•			
SRR	Secondary receiver ready			•			
LL	Local loopback				•		
RL	Remote loopback				•		
TM	Test mode			•			
SS	Select standby				•		
SB	Standby indicator			•			

Primary channel: DM through SI
Secondary channel: SSD through SRR

Centronics interface

Most personal computers use the Centronics parallel data transfer to a printer. The pin connections of the connector, abbreviations and signal descriptions are shown.

All signals are standard TTL, although not all signals necessarily exist in any given interface.

Pin number	Abbreviation	Signal description
1	STROBE	Strobe
2	DATA1	Data line 1
3	DATA2	Data line 2
4	DATA3	Data line 3
5	DATA4	Data line 4
6	DATA5	Data line 5
7	DATA6	Data line 6
8	DATA7	Data line 7
9	DATA8	Data line 8
10	ACKNLG	Acknowledge data
11	BUSY	Busy
12	PE	Paper end
13	SLCT	Select printer
14	AUTO FEED XT	Automatic line feed at end of line
15	NC	No connection
16	OV	Logic ground
17	CHASSIS GND	Printer chassis (not necessarily the same as logic ground)
18	NC	No connection
19 to 30	GND	Single ground
31	INIT	Initialize
32	ERROR	Error
33	GND	Signal ground
34	NC	No connection
35	Logic 1	Logic 1
36	SLCT IN	Select input to printer

RS 232C/RS 449/V24 comparison

RS 232C		RS 449		V24	
AB	Signal ground	SG	Signal ground	102	Signal ground
		SC	Send common	102a	DTE common
		RC	Receive common	102b	DCE common
		IS	Terminal in service		
CE	Ring indicator	IC	Incoming call	125	Calling indicator
CD	Data terminal ready	TR	Terminal ready	108/2	Data terminal ready
CC	Data set ready	DM	Data mode	107	Data set ready
BA	Transmitted data	SD	Send data	103	Transmitted data
BB	Receive data	RD	Receive data	104	Received data
DA	Transmitter signal element Timing (DTE source)	TT	Terminal timing	113	Transmitter signal element Timing (DTE source)
DB	Transmitter signal element Timing (DCE source)	ST	Send timing	114	Transmitter signal element Timing (DCE source)
DD	Receiver signal element timing	RT	Receive timing	115	Receiver signal element timing (DCE source)
CA	Request to send	RS	Request to send	105	Request to send
CB	Clear to send	CS	Clear to send	106	Ready for sending
CF	Received line single detector	RR	Receiver ready	109	Data channel received line signal detector

CG	Signal quality detector	SQ	Signal quality	110	Data signal quality detector
		NS	New signal		
		SF	Select frequency	126	Select transmit frequency
CH	Data signal rate selector (DTE source)	SR	Signalling rate selector	111	Data signalling rate selector (DTE source)
CI	Data signal rate selector (DCE source)	SI	Signalling rate indicator	112	Data signalling rate selector (DCE source)
SBA	Secondary transmitted data	SSD	Secondary send data	118	Transmitted backward channel data
SBB	Secondary received data	SRD	Secondary receive data	119	Received backward channel data
SGA	Secondary request to send	SRS	Secondary request to send	120	Transmit backward channel line signal
SCB	Secondary clear to send	SCS	Secondary clear to send	121	Backward channel ready
SCF	Secondary received line signal detector	SRR	Secondary receiver ready	122	Backward channel received line signal detector
		LL	Local loopback	141	Local loopback
		RL	Remote loopback	140	Remote loopback
		TM	Test mode	143	Test indicator
		SS	Select standby	116	Select standby
		SB	Standby indicator	117	Standby indicator

Reference

M/ACOM Greenpar Ltd, catalogue 1993.

24 Broadcasting

24.1 Standard frequency and time transmissions

Frequency (mHz)	Wavelength (m)	Code	Station location	Country	Power (kW)
60 kHz	5000	MSF	Rugby	England	–
–	–	WWVB	Colorado	USA	–
75 kHz	4000	HBG	–	Switzerland	–
77.5 kHz	3871	DCF77	Mainflingen	DDR	–
1.5	200	HD210A	Guayaquil	Ecuador	–
2.5	120	MSF	Rugby	England	0.5
–	–	WWV	Fort Collins	USA	2.5
–	–	WWVH	Kekaha	Hawaii	5
–	–	ZLF	Wellington	New Zealand	–
–	–	RCH	Tashkent	USSR	1
–	–	JJY	–	Japan	–
–	–	ZUO	Olifantsfontein	South Africa	–
3.33	90.09	CHU	Ottawa	Canada	3
3.81	78.7	HD201A	Guayaquil	Ecuador	–
4.5	66.67	VNG	Victoria	Australia	–
4.996	60.05	RWM	Moscow	USSR	5
5	60	MSF	Rugby	England	0.5
–	–	WWVB	Fort Collins	USA	10
–	–	WWVH	Kekaha	Hawaii	10
–	–	ATA	New Delhi	India	–
–	–	LOL	Buenos Aires	Argentina	2
–	–	IBF	Turin	Italy	5
–	–	RCH	Tashkent	USSR	1
–	–	JJY	–	Japan	–
–	–	ZUO	Olifantsfontein	South Africa	–
5.004	59.95	RID	Irkutsk	USSR	1
6.10	49.2	YVTO	Caracas	Venezuela	–
7.335	40.9	CHU	Ottawa	Canada	10
7.5	40	VNG	Lyndhurst	Australia	5

Frequency (mHz)	Wavelength (m)	Code	Station location	Country	Power (kW)
7.6	39.4	HD210A	Guayaquil	Ecuador	–
8	37.5	JJY	–	Japan	–
8.1675	36.73	LQB9	Buenos Aires	Argentina	5
9996	30.01	RWM	Moscow	USSR	–
10	30	MSF	Rugby	England	0.5
–	–	WWVB	Fort Collins	USA	10
–	–	WWVH	Kekaha	Hawaii	10
–	–	BPM	Xian	China	–
–	–	ATA	New Delhi	India	–
–	–	JJY	–	Japan	–
–	–	LOL	Buenos Aires	Argentina	2
–	–	RTA	Novosibirsk	USSR	5
–	–	RCH	Tashkent	USSR	1
10.004	29.99	RID	Irkutsk	USSR	1
12	25	VNG	Lyndhurst	Australia	10
14.67	20.45	CHU	Ottawa	Canada	3
14.996	20.01	RWM	Moscow	USSR	8
15	20	WWVB	Fort Collins	USA	10
–	–	WWVH	Kekaha	Hawaii	10
–	–	LOL	Buenos Aires	Argentina	2
–	–	RTA	Novosibirsk	USSR	5
–	–	BPM	Xian	China	–
–	–	ATA	New Delhi	India	–
–	–	JJY	–	Japan	–
15.004	19.99	RID	Irkutsk	USSR	1
16.384	18.31	–	Allouis	France	2000
15.55	17.09	LQC20	Buenos Aires	Argentina	5
20	15	WWVB	Fort Collins	USA	2.5
100	3	ZUO	Olifantsfontein	South Africa	–

24.2 Standard frequency formats

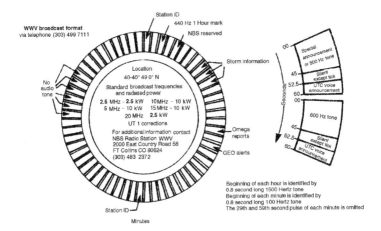

MSF Rugby

Time is inserted in the 60 kHz transmission in two ways, illustrated below.

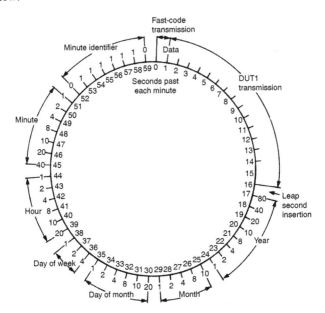

Slow code time and date information is transmitted between the 17th and 59th seconds of the minute-long cycle, in normal BCD coding. *Fast code* time and date BCD coded information is inserted into a 500 ms window in the first second of each minute-long cycle, as illustrated below.

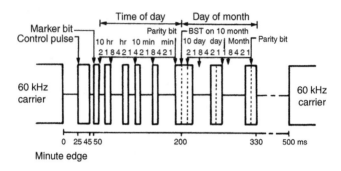

24.3 UK broadcasting bands

Frequency band	Frequency	Use
Long wave	150–285 kHz (2000–1053 m)	AM radio
Medium wave	525–1605 kHz (571–187 m)	AM radio
Band II (VHF)	88–108 MHz	FM radio
Band IV (UHF)	470–582 MHz (channels 21 to 34)	TV
Band V (UHF)	614–854 MHz (channels 39 to 68)	TV
Band VI (SHF)	11.7–12.5 GHz (channels 1 to 40)	Satellite TV

24.4 BBC VHF test tone transmissions

Transmission starts about 4 minutes after the end of Radio 3 programmes on Mondays and Saturdays.

Time min.	Left channel	Right channel	Purpose
–	250 Hz at zero level	440 Hz at zero level	Identification of left and right channels and setting of reference level
2	900 Hz at +7 dB	900 Hz at +7 dB, antiphase to left channel	Adjustment of phase of regenerated subcarrier (see Note 4) and check of distortion with L−R signal only
6	900 Hz at +7 dB	900 Hz +7 dB, in phase with left channel	Check of distortion with L + R signal only
7	900 Hz at +7 dB	No modulation	Check of L to R cross-talk
8	No modulation	900 Hz at +7 dB	Check of R to L cross-talk
9	Tone sequence at −4 dB: 40 Hz 6·3 kHz 100 Hz 10 kHz 500 Hz 12.5 kHz 1000 Hz 14 kHz This sequence is repeated	No modulation	Check of L-channel frequency response and L to R cross-talk at high and low frequencies
11′40″	No modulation	Tone sequence as for left channel	Check of R-channel frequency response and R to L cross-talk at high and low frequencies

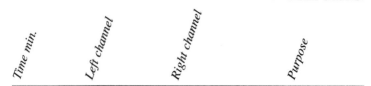

Time min.	Left channel	Right channel	Purpose
14′20″	No modulation	No modulation	Check of noise level in the presence of pilot
15′20″	End of test transmissions		

Notes:

1. This schedule is subject to variation or cancellation to accord with programme requirements and essential transmission tests.
2. The zero level reference corresponds to 40% of the maximum level of modulation applied to either stereophonic channel before pre-emphasis. All tests are transmitted with pre-emphasis.
3. Periods of tone lasting several minutes are interrupted momentarily at one-minute intervals.
4. With receivers having separate controls of subcarrier phase and crosstalk, the correct order of alignment is to adjust first the subcarrier phase to produce maximum output from either the L or the R channel and then to adjust the crosstalk (or 'separation') control for minimum crosstalk between channels.
5. With receivers in which the only control of crosstalk is by adjustment of subcarrier phase, this adjustment of subcarrier phase, this adjustment should be made on the crosstalk checks.
6. Adjustment of the balance control to produce equal loudness from the L and R loudspeakers is best carried out when listening to the announcements during a stereophonic transmission, which are made from a centre-stage position. If this adjustment is attempted during the tone transmissions, the results may be confused because of the occurrence of standing-wave patterns in the listening room.
7. The outputs of most receivers include significant levels of the 19-kHz tone and its harmonics, which may affect signal-level meters. It is important, therefore, to provide filters with adequate loss at these frequencies if instruments are to be used for the above tests.

24.5 Engineering information about broadcast services

Information about all BBC services as well as advice on how best to receive transmissions (including television) can be obtained from:

British Broadcasting Corporation
Engineering Liaison
White City
201 Wood Lane
London W12 7TS

Telephone: 020 8752 5040

Transmitter service maps for most main transmitters can also be supplied, but requests for maps should be accompanied by a stamped addressed A4 sized envelope.

Similarly, information about all IBA broadcast services can be obtained from:

Radio:
Radiocommunications Agency
Wyndham House
189 Marsh Wall
London E14 9SX

Telephone: 020 7211 0211

Television:
The Independent Television Commission
Kings Worthy Court
Kings Worthy
Winchester
Hants SO23 7QA

Telephone: 01962 848647

24.6 Characteristics of UHF terrestrial television systems

24.6.1 World systems

System	Number of lines	Channel width (MHz)	Vision bandwidth (MHz)	Vision/sound separation (MHz)	Vestigial side-band (MHz)	Vision modulation	Sound modulation	Field frequency
A	625	8	5.5	+6	1.25	Neg.	FM	50
B	625	8	5.0	+5.5	1.25	Neg.	FM	50
C	625	8	5.0	+5.5	0.75	Neg.	FM	50
D	625	8	6.0	+6.5	1.25	Pos.	AM	50
E	625	8	6.0	+6.5	0.75	Neg.	FM	50
F	525	6	4.2	+4.5	1.25	Neg.	FM	60

A – UK and Eire
B – Eastern Europe
C – Most of Western Europe, Australia, New Zealand
D – France
E – Russia and Eastern Europe
F – USA, most of Central and South America, Japan

24.6.2 European systems

Country	System	Colour
Austria	C	PAL
Belgium	B	PAL
Bulgaria	No UHF system	
Cyprus	C	PAL
Czechoslovakia	E	SECAM
Denmark	No UHF system	PAL
Finland	C	PAL
France	D	SECAM
Germany	C	PAL

Country	System	Colour
German DR	C	SECAM
Greece	B	PAL
Holland	C	PAL
Hungary	E	SECAM
Iceland	No UHF system	
Ireland	A	PAL
Italy	C	PAL
Luxembourg	D	SECAM
Malta	B	PAL
Monaco	D	SECAM
Norway	C	PAL
Poland	E	SECAM
Portugal	C	PAL
Romania	E	PAL
Spain	C	PAL
Sweden	C	PAL
Switzerland	C	PAL
Turkey	No UHF system	
UK	A	PAL
USSR	E	SECAM
Yugoslavia	B	PAL

24.6.3 UK 625-line television system specification

Channel bandwidth	8 MHz
Upper sideband (vision signal)	5.5 MHz
Lower sideband (vision signal)	1.25 MHz
Vision modulation	AM negative
Sound modulation	FM
Sound deviation (max.)	±50 kHz
Sound pre-emphasis	50 μs
Sound carrier relative to vision carrier	+
Aspect ratio	4:3
Blanking and black level	76%
White level	20% peak
Sync. level	100% peak
Video bandwidth	5.5 MHz

Field frequency	50 Hz
Line frequency	15 625 Hz
Field sync. signal	5 equalizing then 5 broad pulses, followed by 5 equalizing pulses in 7.5 line periods
Field sync. and flyback intervals	2 × 25 line periods
Line period (approx.)	64 μs
Line syn. pulses (approx.)	4.7 μs
Line blanking (approx.)	12 μs
Field sync. pules (broad)	27.3 μs
Field sync. pulses (equalizing)	2.3 μs
Colour subcarrier frequency	4.43361875 MHz
Burst duration	2.25 μs
Burst amplitude	equal to sync
Burst phase	180° ± 45°

24.6.4 UK 625-line television system field blanking details

24.7 Terrestrial television channels

UK

Channel	Frequency (MHz)		Channel	Frequency (MHz)	
	Vision	Sound		Vision	Sound
39	615.25	621.25	54	735.25	741.25
40	623.25	629.25	55	743.25	749.25
41	631.25	637.25	56	751.25	757.25
42	639.25	645.25	57	759.25	765.25
43	647.25	653.25	58	767.25	773.25
44	655.25	661.25	59	775.25	781.25
45	663.25	669.25	60	783.25	789.25
46	671.25	677.25	61	791.25	797.25
47	679.25	685.25	62	799.25	805.25
48	687.25	693.25	63	807.25	813.25
49	695.25	701.25	64	815.25	821.25
50	703.25	709.25	65	823.25	829.25
51	711.25	717.25	66	831.25	837.25
52	719.25	725.25	67	839.25	845.25
53	727.25	733.25	68	847.25	853.25

Republic of Ireland

Channel	Frequency (MHz)		Channel	Frequency (MHz)	
	Vision	Sound		Vision	Sound
IA	45.75	51.75	IF	191.25	197.25
IB	53.75	59.75	IG	199.25	205.25
IC	61.75	67.75	IH	207.25	213.25
ID	175.25	81.25	IJ	215.25	221.25
IE	183.25	189.25			

South Africa

Channel	Frequency (MHz) Vision	Sound	Channel	Frequency (MHz) Vision	Sound
4	175.25	181.25	9	215.25	221.25
5	183.25	189.25	10	223.25	229.25
6	191.25	197.25	11	231.25	237.25
7	199.25	205.25	13	247.43	253.43
8	207.25	213.25			

Australia

Channel	Frequency (MHz) Vision	Sound	Channel	Frequency (MHz) Vision	Sound
0	46.25	51.75	6	175.25	180.75
1	57.25	62.75	7	182.25	187.75
2	64.25	69.75	8	189.25	194.75
3	86.25	91.75	9	196.25	201.75
4	95.25	100.75	10	209.25	214.75
5	102.25	107.75	11	216.25	221.75
5A	138.25	143.75			

New Zealand

Channel	Frequency (MHz) Vision	Sound	Channel	Frequency (MHz) Vision	Sound
1	45.25	50.75	6	189.25	194.75
2	55.25	60.75	7	196.25	201.75
3	62.25	67.75	8	203.25	208.75
4	175.25	180.75	9	210.25	215.75
5	182.25	187.75			

USA

Channel	Frequency (MHz)		Channel	Frequency (MHz)	
	Vision	*Sound*		*Vision*	*Sound*
2	55.25	59.75	43	645.25	649.75
3	61.25	65.75	44	651.25	655.75
4	67.25	71.75	45	657.25	661.75
5	77.25	81.75	46	663.25	667.75
6	83.25	87.75	47	669.25	673.75
7	175.25	179.75	48	675.25	679.75
8	181.25	185.75	49	681.25	685.75
9	187.25	191.75	50	687.25	691.75
10	193.25	197.75	51	693.25	697.75
11	199.25	203.75	52	699.25	703.75
12	205.25	209.75	53	705.25	709.75
13	211.25	215.75	54	711.25	715.75
14	471.25	475.75	55	717.25	721.75
15	477.25	481.75	56	723.25	727.75
16	483.25	487.75	57	729.25	733.75
17	489.25	493.75	58	735.25	739.75
18	495.25	499.75	59	741.25	745.75
19	501.25	505.75	60	747.25	751.75
20	507.25	511.75	61	753.25	757.75
21	513.25	517.75	62	759.25	763.75
22	519.25	523.75	63	765.25	769.75
23	525.25	529.75	64	771.25	775.75
24	531.25	535.75	65	777.25	781.75
25	537.25	541.75	66	783.25	787.75
26	543.25	547.75	67	789.25	793.75
27	549.25	553.75	68	795.25	799.75
28	555.25	559.75	69	801.25	805.75
29	561.25	565.75	70	807.25	811.75
30	567.25	571.75	71	813.25	817.75
31	573.25	577.75	72	819.25	823.75
32	579.25	583.75	73	825.25	829.75
33	585.25	589.75	74	831.25	835.75
34	591.25	595.75	75	837.25	841.75
35	597.25	601.75	76	843.25	847.75
36	603.25	607.75	77	849.25	853.75
37	609.25	613.75	78	855.25	859.75
38	615.25	619.75	79	861.25	865.75
39	621.25	625.75	80	867.25	871.75
40	627.25	631.75	81	873.25	877.75
41	633.25	637.75	82	879.25	883.75
42	639.25	643.75	83	885.25	889.75

24.8 Terrestrial television aerial dimensions

Channel	Dimensions in cm											Channels covered in the UHF groups are:		
UHF Groups	A	B	C	D	E	F	G	H	a	b	c	Group letter	Colour Code	Channels
A	30.1	30	24.1	23	22.8	21.1	20.4	19.9	10.3	10.3	1.8	A	Red	21–34
B	26.5	21.7	18.9	18	17.8	16.5	16	15.5	8.9	8.9	1.8	B	Yellow	39–51
C	23.2	18.2	16	15.3	15	14	13.3	12.2	7.5	7.5	1.8	C	Green	50–66
D	26.1	23.5	18.4	16	15.5	14.8	13.8	13	7.6	7.6	1.8	D	Blue	49–68
E	27	26.5	21.1	18.6	17.9	17.6	16	15.8	15.8	15.8	1.8	E	Brown	39–68

Pattern of general-purpose Yagi array to be used in conjunction with the dimensions given below.

24.9 AM broadcast station classes (USA)

The US AM broadcast band is 540 kHz to 1700 kHz, with 10 kHz channel spacings with centre frequencies divisible by ten (e.g. 780 kHz or 1540 kHz). Other countries in the western hemisphere operate on either 10 kHz channel spacings with frequencies ending in '5', or 9 kHz spacing. The Domestic Class is generally the class of station defined in 47 CFR Section 73.21. The Region 2 Class is generally the class of station as defined in the Region 2 [Western Hemisphere] AM Agreement. This class also corresponds to the class in the 1984 US–Canadian AM Agreement and the 1986 US–Mexican Agreement.

24.9.1 Class A station

A Class A station is an unlimited time station (that is, it can broadcast 24 hours per day) that operates on a clear channel. The operating power shall not be less than 10 kilowatts (kW) or more than 50 kW.

24.9.2 Class B station

A Class B station is an unlimited time station. Class B stations are authorized to operate with a minimum power of 250 watts and a maximum power of 50 kW. (If a Class B station operates with less than 250 W, the RMS must be equal to or greater than 141 mV/m at 1 km for the actual power.) If the station is authorized to operate in the expanded band (1610 to 1700 kHz), the maximum power is 10 kW.

24.9.3 Class C station

A Class C station is an unlimited time station that operates on a local channel. The power shall not be less than 250 W nor more than 1 kW. Class C stations that are licensed to operate with 100 W may continue to operate as licensed.

24.9.4 Class D station

A Class D station operates either daytime, limited time, or unlimited time with a night-time power less than 250 W and an equivalent RMS antenna field less than 141 mV/m at 1 km for the actual power. Class D stations shall operate with daytime powers not less than 0.250 kW nor more than 50 kW. *Note*: If a station is an existing daytime-only station, its class will be Class D.

24.10 FM broadcast frequencies and channel numbers (USA)

From US Code USC 47 CFR 73. The FM broadcast band consists of that portion of the radio frequency spectrum between 88 MHz and 108 MHz. It is divided into 100 channels of 200 kHz each. For convenience, the frequencies available for FM broadcasting (including those assigned to non-commercial educational broadcasting) are given numerical designations which are shown in the table below:

Frequency (MHz)	Channel No.
88.1	201
88.3	202
88.5	203
88.7	204
88.9	205
89.1	206
89.3	207
89.5	208
89.7	209
89.9	210
90.1	211
90.3	212
90.5	213

Frequency (MHz)	Channel No.
90.7	214
90.9	215
91.1	216
91.3	217
91.5	218
91.7	219
91.9	220
92.1	221
92.3	222
92.5	223
92.7	224
92.9	225
93.1	226
93.3	227
93.5	228
93.7	229
93.9	230
94.1	231
94.3	232
94.5	233
94.7	234
94.9	235
95.1	236
95.3	237
95.5	238
95.7	239
95.9	240
96.1	241
96.3	242
96.5	243
96.7	244
96.9	245
97.1	246
97.3	247
97.5	248
97.7	249
97.9	250
98.1	251
98.3	252
98.5	253

Frequency (MHz)	Channel No.
98.7	254
98.9	255
99.1	256
99.3	257
99.5	258
99.7	259
99.9	260
100.1	261
100.3	262
100.5	263
100.7	264
100.9	265
101.1	266
101.3	267
101.5	268
101.7	269
101.9	270
102.1	271
102.3	272
102.5	273
102.7	274
102.9	275
103.1	276
103.3	277
103.5	278
103.7	279
103.9	280
104.1	281
104.3	282
104.5	283
104.7	284
104.9	285
105.1	286
105.3	287
105.5	288
105.7	289
105.9	290
106.1	291
106.3	292
106.5	293

Frequency (MHz)	Channel No.
106.7	294
106.9	295
107.1	296
107.3	297
107.5	298
107.7	299
107.9	300

24.11 US television channel assignments

Channel No.	Frequency (MHz)
1	(No Ch. 1 assigned)
2	54–60
3	60–66
4	66–72
5	76–82
6	82–88
7	174–180
8	180–186
9	186–192
10	192–198
11	198–204
12	204–210
13	210–216
14	470–476
15	476–482
16	482–488
17	488–494
18	494–500
19	500–506
20	506–512
21	512–518
22	518–524
23	524–530
24	530–536
25	536–542
26	542–548
27	548–554

Channel No.	Frequency (MHz)
28	554–560
29	560–566
30	566–572
31	572–578
32	578–584
33	584–590
34	590–596
35	596–602
36	602–608
37	608–614
38	614–620
39	620–626
40	626–632
41	632–638
42	638–644
43	644–650
44	650–656
45	656–662
46	662–668
47	668–674
48	674–680
49	680–686
50	686–692
51	692–698
52	698–704
53	704–710
54	710–716
55	716–722
56	722–728
57	728–734
58	734–740
59	740–746
60	746–752
61	752–758
62	758–764
63	764–770
64	770–776
65	776–782
66	782–788
67	788–794

Channel No.	Frequency (MHz)
68	794–800
69	800–806

Notes:

1. In Alaska, television broadcast stations operating on Channel 5 (76–82 MHz) and on Channel 6 (82–88 MHz) shall not cause harmful interference to and must accept interference from non-Government fixed operations authorized prior to 1 January 1982.
2. Channel 37, 608–614 MHz is reserved exclusively for the radio astronomy service.
3. In Hawaii, the frequency band 488–494 MHz is allocated for non-broadcast use. This frequency band (Channel 17) will not be assigned in Hawaii for use by television broadcast stations.

24.12 License-free bands

In Europe there are a number of license-free bands that can be used by anyone. However, there are restrictions on the use of these bands, both in terms of the application and the transmitted power. In particular, the manufacturer of transmitters operating in these bands must certify the equipment as meeting the required emission limits.

49.82 to 49.98 MHz – general purpose
173.2 to 173.35 MHz – telemetry. Note: 173.225 MHz is for short range alarms only
433.05 to 434.79 MHz – telemetry and vehicle security
458.5 to 458.95 MHz – industrial or commercial telemetry
868 to 870 MHz – general purpose.

Worldwide agreement has resulted in two bands, at 2.4 GHz and 5 GHz, being allocated to wireless LANs. Both of these bands are for industrial, scientific and medical (ISM) use.

The 2.4 GHz allocation covers 2.4 to 2.4835 GHz. Restrictions on the use of this ISM band include the use of carrier frequency hopping in step multiples of 1 MHz. This band is used by wireless LANs operating under IEEE802.11B and Bluetooth.

The IEEE802.11B standard uses 5 MHz carrier frequency steps, with some international variations. Europe allows the use of 13 channels (channel 1 to 13, with centre frequencies 2412 MHz to 2472 MHz), but the USA only allows the use of 11 channels (channel 1 to 11, with centre frequencies 2412 MHz to 2462 MHz). In Japan only channel 14 is allowed (no hopping), using a carrier centre frequency of 2477 MHz.

Bluetooth uses 79 hop frequencies, with multiples of a 1 MHz step size. The lowest carrier frequency is at 2402 MHz (channel 0) and the highest is 2480 MHz (channel 78). The frequency hopping pattern is psuedo-random, with 1600 hops per second.

The 5 GHz allocation is actually two sub-bands, one covering 5.15 to 5.35 GHz and the other covering 5.47 GHz to 5.725 GHz. The higher frequency ISM band allows 1 watt radiated power, rather than the 200 mW limit of the lower frequency band. As with the 2.4 GHz ISM band, sources in the 5 GHz band must use carrier frequency hopping with 5 MHz steps.

24.13 Calculating radio antenna great circle bearings

Aiming radio antennas to target a particular area of the world requires calculation of the *great circle bearing* between your location and the other stations' location. That bearing is calculated from some simple spherical trigonometry using a hand-held calculator or a computer program. Before talking about the maths, however, we need to establish a frame of reference that makes the system work.

24.13.1 Latitude and longitude

The need for navigation on the surface of the Earth caused the creation of a grid system uniquely to locate points on the surface of our globe. *Longitude* lines run from the north pole to the south pole, i.e. from north to south.

The reference point (longitude zero), called the *prime meridian*, runs through Greenwich, England. The longitude of the prime meridian is 0 degrees. Longitudes west of the prime meridian are given a plus sign (+), while longitudes east of the prime are given a minus (−) sign. If you continue the prime meridian through the poles to the other side of the Earth it has a longitude of 180 degrees. Thus, the longitude values run from −180 degrees to +180 degrees, with ±180 degrees being the same line.

The observatory at Greenwich is also the point against which relative time is measured. Every 15 degree change of longitude is equivalent to a one hour difference with the Greenwich time. To the west, subtract one hour for each 15 degrees and to the east add one

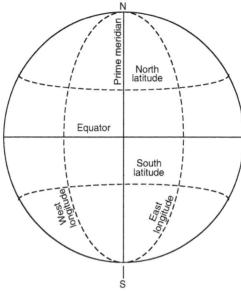

Figure 24.1 Lines of longitude and latitude

hour for each 15 degrees. Thus, the time on the East Coast of the United States is −5 hours relative to Greenwich time. At one time, we called time along the prime meridian *Greenwich mean time* (GMT), also called *Zulu time* to simplify matters for CW operators.

Latitude lines are measured against the equator, with distances north of the equator being taken as positive, and distances south of the equator being negative. The equator is 0 degrees latitude, while the north pole is +90 degrees latitude and the south pole is −90 degrees latitude.

Long ago navigators learned that the latitude can be measured by 'shooting' the stars and consulting a special atlas to compare the angle of certain stars with tables that translate to latitude numbers. The longitude measurement, however, is a bit different. For centuries sailors could measure latitude, but had to guess longitude (often with tragic results). In the early eighteenth century, the British government offered a large cash prize to anyone who could design a chronometer that could be taken to sea. By keeping the chronometer set accurately to Greenwich mean time, and comparing GMT against local time (i.e. at a time like high noon when the position of the sun is easy to judge), the longitude could be calculated. If you are interested in this subject, then most decent libraries have books on celestial navigation.

304

24.13.2 The great circle

On the surface of a globe, a curved line called a *great circle path* is the shortest distance between two points.

Consider two points on a globe: 'A' is your location, while 'B' is the other station's location. The distance 'D' is the great circle path between 'A' and 'B'.

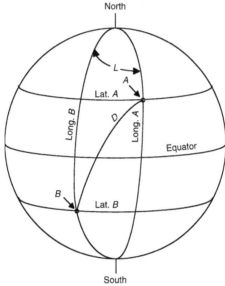

Figure 24.2 Great circle path

The great circle path length can be expressed in either degrees or distance (e.g. miles, nautical miles or kilometres). To calculate the distance, it is necessary to find the difference in longitude (L) between your longitude (LA) and the other station's longitude (LB): $L = LA - LB$. Keep the signs straight. For example, if your longitude (LA) is 40 degrees, and the other station's longitude (LB) is −120 degrees, then $L = 40 - (-120) = 40 + 120 = 160$. The equation for distance (D) is:

$$\cos D = (\sin A \times \sin B) + (\cos A \times \cos B \times \cos L)$$

where:
D   is the angular great circle distance
A   is your latitude
B   is the other station's latitude.

To find the actual angle, take the arccos of the above equation, i.e.

$$D = \arccos(\cos D)$$

In the next equation you will want to use D in angular measure, but later on will want to convert D to miles. To do this, multiply D in degrees by 69.4. Or, if you prefer metric measures, then $D \times 111.2$ yields kilometres. This is the approximate distance in statute miles between 'A' and 'B'.

To find the bearing from true north, work the equation below:

$$C = \arccos\left[\frac{\sin B - (\sin A \times \cos D)}{(\cos A \times \sin D)}\right]$$

However, this equation won't always give you the right answer unless you make some corrections.

The first problem is the 'same longitude error', i.e. when both stations are on the same longitude line. In this case, $L = LA - LB = 0$. If LAT $A >$ LAT B, then $C = 180$ degrees, but if LAT $A <$ LAT B, then $C = 0$ degrees. If LAT $A =$ LAT B, then what's the point of all these calculations?

The next problem is found when the condition $-180° \leq L \leq +180°$ is not met, i.e. when the absolute value of L is greater than $180°$, $\text{ABS}(L) > 180°$. In this case, either add or subtract 360 in order to make the value between $\pm180°$:

If $L > +180$, then $L = L - 360$

If $L < -180$, then $L = L + 360$

One problem seen while calculating these values on a computer or hand calculator is the fact that the $\sin(X)$ and $\cos(X)$ cover different ranges. The $\sin(X)$ function returns values from $0°$ to $360°$, while the $\cos(X)$ function returns values only over $0°$ to $180°$. If L is positive, then the result bearing C is accurate, but if L is negative then the actual value of $C = 360 - C$. The following test is necessary:

If $L < 0$ then

$L = 360 - L$

Else $L = L$

End if

Another problem is seen whenever either station is in a high latitude near either pole ($\pm 90°$), or where both locations are very close together, or where the two locations are antipodal (i.e. on opposite points on the Earth's surface). The best way to handle these problems is to use a different equation that multiplies by the cosecant of D (i.e. $\operatorname{cosec}(D)$), rather than dividing by sine of D (i.e. $\sin(D)$).

25 Abbreviations and symbols

25.1 Abbreviations

Many abbreviations are found as either capital or lower case letters, depending on publishers' styles. Symbols should generally be standard, as shown.

A	Ampere or anode
ABR	Auxiliary bass radiator
a.c.	Alternating current
Ack	Acknowledgement
A/D	Analogue to digital
ADC	Analogue to digital converter
Ae	Aerial
a.f.	Audio frequency
a.f.c.	Automatic frequency control
a.g.c.	Automatic gain control
a.m.	Amplitude modulation
AMPS	Advanced mobile phone system
ANSI	American National Standards Institute
ASA	Acoustical Society of America
ASCII	American Standard Code for Information Interchange
a.t.u.	Aerial tuning unit
AUX	Auxiliary
a.v.c.	Automatic volume control
AWG	American Wine Gauge
b	Base of transistor
BAF	Bonded acetate fibre
B & S	Brown & Sharpe (U.S.) wire gauge
b.p.s.	Bits per second
BR	Bass reflex
BSI	British Standards Institution
BW	Bandwidth
C	Capacitor, cathode, centigrade, coulomb
c	Collector of transistor, speed of light
CB	Citizen's band
CCD	Charge coupled device
CCIR	International Radio Consultative Committee

CCITT	International Telegraph and Telephone Consultative Committee
CCTV	Closed circuit television
CDMA	Code division multiple access
chps	Characters per second
CISPR	Comité International Special Des Peturbations (radio interference standards body)
CLK	Clock signal
CMOS	Complementary metal oxide semiconductor
CrO_2	Chromium dioxide
CPU	Central processor unit
CTCSS	Continuous tone controlled signalled system
CTD	Charge transfer device
c.w.	Continuous wave
D	Diode
d	Drain of an f.e.t.
D/A	Digital to analogue
DAC	Digital to analogue converter
dB	Decibel
d.c.	Direct current
DCC	Double cotton covered
DCE	Data circuit-terminating equipment
DCS	Digital Communications Service
DF	Direction finding
DIL	Dual-in-line
DIN	German standards institute
DMA	Direct memory access
DPDT	Double pole, double throw
DPSK	Differential phase shift keying
DPST	Double pole, single throw
dsb	or dsbam. Double sideband amplitude modulation
DSRR	Digital short range radio
DTE	Data terminal equipment
DTL	Diode-transistor logic
DTMF	Dual tone multi-frequency
DX	Long distance reception
e	Emitter of transistor
EAROM	Electrically alterable read only memory
ECL	Emitter coupled logic
EDGE	Enhanced Data for GSM Evolution
e.h.t.	Extremely high tension (voltage)

EIRP	Effective Isotropic Radiated Power
e.m.f.	Electromotive force
en	Enamelled
EPROM	Erasable programmable read only memory
EQ	Equalization
EROM	Erasable read only memory
ERP	Effective radiated power
ETACS	Extended total access communications system
F	Farad, fahrenheit or force
f	Frequency
FDD	Frequency Division Duplex
FDM	Frequency division multiplex
FDMA	Frequency division multiple access
Fe	Ferrous
FeCr	Ferri-chrome
f.e.t.	Field effect transistor
FFSK	Fast frequency shift keying
f.m.	Frequency modulation
f.r.	Frequency response or range
f.s.d.	Full-scale deflection
FSK	Frequency shift keying
G	Giga (10^9)
g	Grid, gravitational constant
GMSK	Gaussian minimum shift keying
GSM	Global system mobile
H	Henry
h.f.	High frequency
Hz	Hertz (cycles per second)
I	Current
IB	Infinite baffle
i.c.	Integrated circuit
IF	Intermediate frequency
IHF	Institute of High Fidelity (U.S.)
$I^2L(HL)$	Integrated injection logic
i.m.d.	Intermodulation distortion
i/p	Input
i.p.s.	Inches per second
K	Kilo, in computing terms ($= 2^{10} = 1024$), or degrees Kelvin
k	Kilo (10^3) or cathode
L	Inductance or lumens
l.e.d.	Light emitting diode

l.f.	Low frequency
LIN	Linear
LOG	Logarithmic
LS	Loudspeaker
LSI	Large scale integration
l.w.	Long wave (approx. 1100–2000 m)
M	Mega (10^6)
m	Milli (10^{-3}) or metres
m.c.	Moving coil
MHz	Megahertz
mic	Microphone
MOS	Metal oxide semiconductor
MPU	Microprocessor unit
MPX	Multiplex
MSC	Mobile Switching Centre
MSK	Minimum shift keying
m.w.	Medium wave (approx. 185–560 m)
n	Nano (10^9)
NAB	National Association of Broadcasters
Ni-Cad	Nickel-cadmium
n/c	Not connected; normally closed
n/o	Normally open
NMOS	Negative channel metal oxide semiconductor
NRZ	Non-return to zero
o/c	Open channel; open circuit
o/p	Output
op-amp	Operational amplifier
p	Pico (10^{-12})
PA	Public address
PABX	Private automatic branch exchange
PAL	Phase alternation, line
p.a.m.	Pulse amplitude modulation
PCB	Printed circuit board
PCM	Pulse code modulation
PCN	Personal communications network
PCS	Personal communication system
PLA	Programmable logic array
PLL	Phase locked loop
pm	Phase modulation
PMOS	Positive channel metal oxide semiconductor
P.P.M.	Peak programme meter
p.r.f.	Pulse repetition frequency

PROM	Programmable read only memory
PSK	Phase shift keying
PSS	Packet SwitchStream
PSTN	Public Switched Telephone Network
PSU	Power supply unit
PTFE	Polytetrafluoroethylene
PU	Pickup
PUJT	Programmable unijunction transistor
Q	Quality factor; efficiency of tuned circuit, charge
QAM	Quadrature (or quaternary) amplitude modulation
QPSK	Quadrature (or quaternary) phase shift keying
R	Resistance
RAM	Random access memory
RCF	Recommended crossover frequency
r.f.	Radio frequency
r.f.c.	Radio frequency choke (coil)
RFI	Radio frequency interference
RIAA	Record Industry Association of America
r.m.s.	Root mean square
ROM	Read only memory
RTL	Resistor transistor logic
R/W	Read/write
RX	Receiver
S	Siemens
s	Source of an f.e.t.
s/c	Short circuit
SCR	Silicon-controlled rectifier
s.h.f.	Super high frequency
SI	International system of units
S/N	Signal-to-noise.
SPL	Sound pressure level
SPDT	Single pole, double throw
SPST	Single pole, single throw
ssb	Single sideband amplitude modulation
ssbdc	Single sideband diminished carrier
ssbsc	Single sideband suppressed carrier
SSI	Small scale integration
s.w.	Short wave (approx. 10–60 m)
s.w.g.	Standard wire gauge
s.w.r.	Standing wave ratio
T	Tesla

312

TACS	Total access communications system
TDD	Time division duplex
TDM	Time division multiplex
TDMA	Time division multiple access
t.h.d.	Total harmonic distortion
t.i.d.	Transient intermodulation distortion
TR	Transformer
t.r.f.	Tuned radio frequency
TRS	Transmitter repeater station
TTL	Transistor transistor logic
TTY	Teletype unit
TVI	Television interface; television interference
TX	Transmitter
UART	Universal asynchronous receiver transmitter
u.h.f.	Ultra high frequency (approx. 470–854 MHz)
u.j.t.	Unijunction transistor
ULA	Uncommitted logic array
V	Volts
VA	Volt-amps
v.c.a.	Voltage controlled amplifier
v.c.o.	Voltage controlled oscillator
VCT	Voltage to current transactor
v.h.f.	Very high frequency (approx. 88–216 MHz)
v.l.f.	Very low frequency
VSB	Vestigial side band
VSWR	Voltage standing wave ratio
VU	Volume unit
W	Watts
WAP	Wireless application protocol
Wb	Weber
W/F	Wow and flutter
WML	Wireless mark-up language
w.p.m.	Words per minute
X	Reactance
Xtal	Crystal
Z	Impedance
ZD	Zener diode

25.2 Letter symbols by unit name

Unit	Symbol	Notes
ampere	A	SI unit of electric current.
ampere (turn)	At	SI unit of magnetomotive force.
ampere-hour	Ah	
ampere per metre	$A\,m^{-1}$	SI unit of magnetic field strength.
angstrom	Å	$1\,Å = 10^{-10}\,m$.
apostilb	asb	$1\,asb(1/\pi)cd\,m^{-2}$ A unit of luminance. The SI unit, candela per square metre, is preferred.
atmosphere:		
standard atmosphere	atm	$1\,atm = 101\,325\,N\,m^{-2}$.
technical atmosphere	at	$1\,at = 1\,kgf\,cm^{-2}$.
atomic mass unit (unified)	u	The (unified) atomic mass unit is defined as one-twelfth of the mass of an atom of the ^{12}C nuclide. Use of the old atomic mass unit (amu), defined by reference to oxygen, is deprecated.
bam	b	$1\,b10^{-28}\,m^2$.
bar	bar	$1\,bar = 100\,000\,N\,m^{-2}$.
baud	Bd	Unit of signalling speed equal to one element per second.
becquerel	Bq	$1\,Bq = 1\,s^{-1}$. SI unit of radioactivity.
bel	B	
bit	b	
British thermal unit	Btu	

Unit	Symbol	Notes
calorie (International Table calorie)	cal_{IT}	$1\,cal° - = 4.1868\,J$. The 9th Conférence Générale des Poids et Mesures adopted the joule as the unit of heat, avoiding the use of the calorie as far as possible.
calorie (thermochemical calorie)	cal	$1\,cal = 4.1840\,J$. (See note for International Table calorie.)
candela	cd	SI unit of luminous intensity.
candela per square inch	$cd\,in^{-2}$	Use of the SI unit, candela per square metre, is preferred.
candela per square metre	$cd\,m^{-2}$	SI unit of luminance. The name nit has been used.
candle		The unit of luminous intensity has been given the name *candela*; use of the word *candle* for this purpose is deprecated.
centimetre	cm	
circular mil	cmil	$1\,cmil = (\pi/4)10^{-6}\,in^2$.
coulomb	C	SI unit of electrical charge.
cubic centimetre	cm^3	
cubic foot	ft^3	
cubic foot per minute	$ft^3\,min^{-1}$	
cubic foot per second	$ft^3\,s^{-1}$	
cubic inch	in^3	
cubic metre	m^3	
cubic metre per second	$m^3\,s^{-1}$	
cubic yard	yd^3	
curie	Ci	Unit of activity in the field of radiation dosimetry.
cycle	c	
cycle per second	$c\,s^{-1}$	Deprecated. Use hertz.
decibel	dB	
degree (plane angle)	°	

Unit	Symbol	Notes
degree (temperature):		Note that there is no space
degree Celsius	°C	between the symbol ° and
degree Fahrenheit	°F	the letter. The use of the
		word *centigrade* of the
		Celsius temperature scale
		was abandoned by the
		Conférence Générale des
		Poids et Mesures in 1948.
degree Kelvin		See Kelvin.
degree Rankine	°R	
dyne	dyn	
electronvolt	eV	
erg	erg	
erlang	E	Unit of telephone traffic.
farad	F	SI unit of capacitance.
foot	ft	
footcandle	fc	Use of the SI unit of illuminance, the lux (lumen per square metre), is preferred.
footlambert	fL	Use of the SI unit, the candela per square metre, is preferred.
foot per minute	ft min^{-1}	
foot per second	ft s^{-1}	
foot per second squared	ft s^{-2}	
foot pound-force	ft lb$_f$	
gal	Gal	$1\,\text{Gal} = 1\,\text{cm s}^{-2}$.
gallon	gal	The gallon, quart, and pint differ in the US and the UK, and their use is deprecated.
gauss	G	The gauss is the electromagnetic CGS (Centimetre Gram Second) unit of magnetic flux density. The SI unit, tesla, is preferred.
gigaelectronvolt	GeV	

Unit	Symbol	Notes
gigahertz	GHz	
gilbert	Gb	The gilbert is the electromagnetic CGS (Centimetre Gram Second) unit of magnetomotive force. Use of the SI unit, the ampere (or ampere-turn), is preferred.
grain	gr	
gram	g	
gray	Gy	$1\ Gy = 1\ J\ kg^{-1}$. SI unit of absorbed dose.
henry	H	
hertz	Hz	SI unit of frequency.
horsepower	hp	Use of the SI unit, the watt, is preferred.
hour	h	Time may be designated as in the following example; $9^h46^m30^s$.
inch	in	
inch per second	in s^{-1}	
joule	J	SI unit of energy.
joule per Kelvin	J K^{-1}	SI unit of heat capacity and entropy.
Kelvin	K	SI unit of temperature (formerly called *degree Kelvin*). The symbol K is now used without the symbol°.
kiloelectronvolt	KeV	
kilogauss	kG	
kilogram	kg	SI unit of mass.
kilogram-force	kg$_f$	In some countries the name *kilopond* (kp) has been adopted for this unit.
kilohertz	kHz	
kilohm	kΩ	
kilojoule	kJ	
kilometre	km	

Unit	Symbol	Notes
kilometre per hour	km h^{-1}	
kilopond	kp	See kilogram-force.
kilovar	kvar	
kilovolt	kV	
kilovoltampere	kVA	
kilowatt	kW	
kilowatthour	kWh	
knot	kn	1 kn = 1 nmi h^{-1}.
lambert	L	The lambert is the CGS (Centimetre Gram Second) unit of luminance. The SI unit, candela per square metre, is preferred.
litre	l	
litre per second	l s^{-1}	
lumen	lm	SI unit of luminous flux.
lumen per square foot	lm ft^{-2}	Use of the SI unit, the lumen per square metre, is preferred.
lumen per square metre	lm m^{-2}	SI unit of luminous excitance.
lumen per watt	lm W^{-1}	SI unit of luminous efficacy.
lumen second	lm s	SI unit of quantity of light.
lux	lx	1 lx = 1 lm m^{-2}. SI unit of illuminance.
maxwell	Mx	The maxwell is the electromagnetic CGS (Centimetre Gram Second) unit of magnetic flux. Use of the SI unit, the weber, is preferred.
megaelectronvolt	MeV	
megahertz	MHz	
megavolt	MV	
megawatt	MW	
megohm	MΩ	
metre	m	SI unit of length.
mho	mho	1 mho = 1 Ω^{-1} = 1 S.
microampere	μA	
microbar	μbar	

Unit	Symbol	Notes
microfarad	μF	
microgram	μg	
microhenry	μH	
micrometre	μm	
micron		The name *micrometre* (μm) is preferred.
microsecond	μS	
microwatt	μW	
mil	mil	1 mil = −0.001 in.
mile		
nautical	nmi	
statute	mi	
mile per hour	mi h^{-1}	
milliampere	mA	
millibar	mbar	mb may be used.
milligal	mGal	
milligram	mg	
millihenry	mH	
millilitre	ml	
millimetre	mm	
conventional millimetre of mercury	mm Hg	1 mm Hg = 133.322 N m^{-2}.
millimicron		The name *nanometre* (nm) is preferred.
millisecond	ms	
millivolt	mV	
milliwatt	mW	
minute (plane angle)	...′	
minute (time)	min	Time may be designated as in the following example: 9^{h}46^{m}30^s.
mole	mol	SI unit of amount of substance.
nanoampere	nA	
nanofarad	nF	
nanometre	nm	
nanosecond	ns	
nanowatt	nW	
nautical mile	nmi	

Unit	Symbol	Notes
neper	Np	
newton	N	SI unit of force.
newton metre	Nm	
newton per square metre	$N\,m^{-2}$	See pascal.
nit	nt	$1\,nt = 1\,cd\,m^{-2}$. See candela per square metre.
oersted	Oe	The oersted is the electromagnetic CGS (Centimetre Gram Second) unit of magnetic field strength. Use of the SI unit, the ampere per metre, is preferred.
ohm	Ω	SI unit of electrical resistance.
ounce (avoirdupois)	oz	
pascal	Pa	SI unit of pressure or stress. $1\,Pa = 1\,N\,m^{-2}$.
picoampere	pA	
picofarad	pF	
picosecond	ps	
picowatt	pW	
pint	pt	The gallon, quart, and pint differ in the US and the UK, and their use is deprecated.
pound	lb	
poundal	pdl	
pound-force	lb_f	
pound-force feet	lb_f ft	
pound-force per square inch	$lb_f\,in^{-2}$	
pound per square inch		Although use of the abbreviation psi is common, it is not recommended. See pound-force per square inch.

Unit	Symbol	Notes
Quart	qt	The gallon, quart, and pint differ in the US and the UK, and their use is deprecated.
rad	rd	Unit of absorbed dose in the field of radiation dosimetry.
revolution per minute	r min^{-1}	Although use of the abbreviation rpm is common, it is not recommended.
revolution per second	r s^{-1}	
roentgen	R	Unit of exposure in the field of radiation dosimetry.
second (plane angle)	$\ldots''$	
second (time)	s	SI unit of time. Time may be designated as in the following example: $9^h46^m30^s$.
siemens	S	SI unit of conductance. $1\,S = 1\,\Omega^{-1}$.
square foot	ft^2	
square inch	in^2	
square metre	m^2	
square yard	yd^2	
steradian	sr	SI unit of solid angle.
stilb	sb	$1\,sb = 1\,cd\ cm^{-2}$. A CGS unit of luminance. Use of the SI unit, the candela per square metre, is preferred.
tesla	T	SI unit of magnetic flux density. $1\,T = 1\,Wb\ m^{-2}$.
tonne	t	$1\,t = 1000\,kg$.
(unified) atomic mass unit	u	See atomic mass unit (unified).
var	var	Unit of reactive power.
volt	V	SI unit of electromotive force.
voltampere	VA	SI unit of apparent power.
watt	W	SI unit of power.
watthour	Wh	

Unit	Symbol	Notes
watt per steradian	W sr^{-1}	SI unit of radiant intensity.
watt per steradian square metre	W (sr m^2)$^{-1}$	SI unit of radiance.
weber	Wb	SI unit of magnetic flux. 1 Wb = 1 V s.
yard	yd	

25.3 Electric quantities

Quantity	Symbol	Unit	Symbol
Admittance	Y	siemens	S
Angular frequency	ω	hertz	Hz
Apparent power	S	watt	W
Capacitance	C	farad	F
Charge	Q	coulomb	C
Charge density	ρ	coulomb per square metre	C m^{-2}
Conductance	G	siemens	S
Conductivity	κ, γ, σ	siemens per metre	S m^{-1}
Current	I	ampere	A
Current density	j, J	ampere per square metre	A m^{-2}
Displacement	D	coulomb per square metre	C m^{-2}
Electromotive force	E	volt	V
Energy	E	joule	J
Faraday constant	F	coloumb per mole	C mol^{-1}
Field strength	E	volt per metre	V m^{-1}
Flux	ψ	coulomb	C
Frequency	v, f	hertz	Hz
Impedance	Z	ohm	Ω
Light, velocity of in a vacuum	c	metre per second	m s^{-1}
Period	T	second	s
Permeability	μ	henry per metre	H m^{-1}
Permeability of space	μ_o	henry per metre	H m^{-1}
Permeance	Λ	henry	H

Quantity	Symbol	Unit	Symbol
Permittivity	ε	farad per metre	$\mathrm{F\,m^{-1}}$
Permittivity of space	ε_0	farad per metre	$\mathrm{F\,m^{-1}}$
Phase	ϕ	–	–
Potential	V, U	volt	V
Power	P	watt	W
Quality factor	Q	–	–
Reactance	X	ohm	Ω
Reactive power	Q	watt	W
Relative permeability	μ_r	–	–
Relative permittivity	ε_r	–	–
Relaxation time	τ	second	s
Reluctance	R	reciprocal henry	$\mathrm{H^{-1}}$
Resistance	R	ohm	Ω
Resistivity	ρ	ohm metre	Ω m
Susceptance	B	siemens	S
Thermodynamic temperature	T	kelvin	K
Time constant	τ	second	s
Wavelength	λ	metre	m

26 Miscellaneous data

26.1 Fundamental constants

Constant	Symbol	Value
Boltzmann constant	k	$1.38062 \times 10^{-23} \text{ J K}^{-1}$
Electron charge, proton charge	e	$\pm 1.60219 \times 10^{-19} \text{ C}$
Electron charge-to-mass ratio	e/m	$1.7588 \times 10^{11} \text{ C kg}^{-1}$
Electron mass	m_e	$9.10956 \times 10^{-31} \text{ kg}$
Electron radius	r_e	$2.81794 \times 10^{-15} \text{ m}$
Faraday constant	F	$9.64867 \times 10^{4} \text{ C mol}^{-1}$
Neutron mass	m_n	$1.67492 \times 10^{-27} \text{ kg}$
Permeability of space	μ_o	$4\pi \times 10^{-7} \text{ H m}^{-1}$
Permittivity of space	ε_o	$8.85419 \times 10^{-12} \text{ F m}^{-1}$
Planck constant	h	$6.6262 \times 10^{-34} \text{ J s}$
Proton mass	m_p	$1.67251 \times 10^{-27} \text{ kg}$
Velocity of light	c	$2.99793 \times 10^{8} \text{ m s}^{-1}$

26.2 Electrical relationships

Amperes $\times$ ohms = **volts**
Volts $\div$ amperes = **ohms**
Volts $\div$ ohms = **amperes**
Amperes $\times$ volts = **watts**
(Amperes)^2 $\times$ ohms = **watts**
(Volts)^2 $\div$ ohms = **watts**
Joules per second = **watts**
Coulombs per second = **amperes**
Amperes $\times$ seconds = **coulombs**
Farads $\times$ volts = **coulombs**
Coulombs $\div$ volts = **farads**
Coulombs $\div$ farads = **volts**
Volts $\times$ coulombs = **joules**
Farads $\times$ (volts)^2 = **joules**

26.3 Dimensions of physical properties

Length: metre [L]. Mass: kilogram [M]. Time: second [T]. Quantity of electricity: coulomb [Q]. Area: square metre [L^2]. Volume: cubic metre [L^3].

Velocity: metre per second	$[LT^{-1}]$
Acceleration: metre per second^2	$[LT^{-2}]$
Force: newton	$[MLT^{-2}]$
Work: joule	$[ML^2T^{-2}]$
Power: watt	$[ML^2T^{-3}]$
Electric current: ampere	$[QT^{-1}]$
Voltage: volt	$[ML^2T^{-2}Q^{-1}]$
Electric resistance: ohm	$[ML^2T^{-1}Q^{-2}]$
Electric conductance: siemens	$[M^{-1}L^{-2}TQ^2]$
Inductance: henry	$[ML^2Q^{-2}]$
Capacitance: farad	$[M^{-1}L^{-2}T^2Q^2]$
Current density: ampere per metre^2	$[L^{-2}T^{-1}Q]$
Electric field strength: volt per metre	$[MLT^{-2}Q^{-1}]$
Magnetic flux: weber	$[MLT^2T^{-1}Q^{-1}]$
Magnetic flux density: tesla	$[MT^{-1}Q^{-1}]$
Energy: joule	$[ML^2T^{-2}]$
Frequency: hertz	$[T^{-1}]$
Pressure: pascal	$[ML^{-1}T^{-2}]$

26.4 Fundamental units

Quantity	Unit	Symbol
Amount of a substance	mole	mol
Charge	coulomb	C
Length	metre	m
Luminous intensity	candela	cd
Mass	kilogram	kg
Plane angle	radian	rad
Solid angle	steradian	sr
Thermodynamic temperature	kelvin	K
Time	second	s

26.5 Greek alphabet

Capital letters	Small letters	Greek name	English equivalent	Capital letters	Small letters	Greek name	English equivalent
A	α	Alpha	a	N	ν	Nu	n
B	β	Beta	b	Ξ	ξ	Xi	x
Γ	γ	Gamma	g	O	o	Omicron	$\breve{o}$
Δ	δ	Delta	d	Π	π	Pi	p
E	ε	Epsilon	e	P	ρ	Rho	r
Z	ζ	Zeta	z	Σ	σ	Sigma	s
H	η	Eta	é	T	τ	Tau	t
Θ	θ	Theta	th	Y	υ	Upsilon	u
I	ι	Iota	i	Φ	ϕ	Phi	ph
K	κ	Kappa	k	X	χ	Chi	ch
Λ	λ	Lambda	l	Ψ	ψ	Psi	ps
M	μ	Mu	m	Ω	ω	Omega	ö

26.6 Standard units

Ampere Unit of electric current, the constant current which, if maintained in two straight parallel conductors of infinite length of negligible circular cross-section and placed one metre apart in a vacuum, will produce between them a force equal to 2×10^{-7} newton per metre length.

Ampere-hour Unit of quantity of electricity equal to 3 600 coulombs. One unit is represented by one ampere flowing for one hour.

Candela Unit of luminous intensity. It is the luminous intensity, in the perpendicular direction, of a surface of $1/600\,000\,\text{m}^{-2}$ of a full radiator at the temperature of freezing platinum under a pressure of 101 325 newtons m^{-2}.

Coulomb Unit of electric charge, the quantity of electricity transported in one second by one ampere.

Decibel (dB) Unit of acoustical or electrical power ratio. Although the bel is officially the unit, this is usually regarded as being too large, so the decibel is preferred. The difference between two power levels is P_1 and P_2, is given as

$$10 \log_{10} \frac{P_1}{P_2} \text{ decibels}$$

Farad Unit of electric capacitance. The capacitance of a capacitor between the plates of which there appears a difference of potential of one volt when it is charged by one coulomb of electricity. Practical units are the microfarad (10^{-6} farad), the nanofarad (10^{-9} farad) and the picofarad (10^{-12} farad).

Henry Unit of electrical inductance. The inductance of a closed circuit in which an electromotive force of one volt is produced when the electric current in the circuit varies uniformly at the rate of one ampere per second. Practical units are the microhenry (10^{-6} henry) and the millihenry (10^{-3} henry).

Hertz Unit of frequency. The number of repetitions of a regular occurrence in one second.

Joule Unit of energy, including work and quantity of heat. The work done when the point of application of a force of one newton is displaced through a distance of one metre in the direction of the force.

Kilovolt-ampere 1000 volt-amperes.

Kilowatt 1000 watts.

Light, velocity of Light waves travel at 300 000 kilometres per second (approximately). Also the velocity of radio waves.

Lumen m^{-2}, lux Unit of illuminance of a surface.

Mho Unit of conductance, see Siemens.

Newton Unit of force. That force which, applied to a mass of one kilogram, gives it an acceleration of one metre per second per second.

Ohm Unit of electric resistance. The resistance between two points of a conductor when a constant difference of potential of one volt, applied between these two points, produces in the conductor a current of one ampere.

Pascal Unit of sound pressure. Pressure is usually quoted as the root mean square pressure for a pure sinusoidal wave.

Siemens Unit of conductance, the reciprocal of the ohm. A body having a resistance of 4 ohms would have a conductance of 0.25 siemens.

Sound, velocity of Sound waves travel at 332 metres per second in air (approximately) at sea level.

Tesla Unit of magnetic flux density, equal to one weber per square metre of circuit area.

Volt Unit of electric potential. The difference of electric potential between two points of a conducting wire carrying a constant current of one ampere, when the power dissipated between these points is equal to one watt.

Volt-ampere The product of the root-mean-square volts and root-mean-square amperes.

Watt Unit of power, equal to one joule per second. Volts times amperes equals watts.

Weber Unit of magnetic flux. The magnetic flux which, linking a circuit of one turn, produces in it an electromotive force of one volt as it is reduced to zero at a uniform rate in one second.

26.7 Decimal multipliers

Prefix	Symbol	Multiplier	Prefix	Symbol	Multiplier
tera	T	10^{12}	centi	c	10^{-2}
giga	G	10^{9}	milli	m	10^{-3}
mega	M	10^{6}	micro	μ	10^{-6}
kilo	k	10^{3}	nano	n	10^{-9}
hecto	h	10^{2}	pico	p	10^{-12}
deka	da	10	femto	f	10^{-15}
deci	d	10^{-1}	atto	a	10^{-18}

26.8 Useful formulae

Boolean Algebra (laws of)

Absorption:
$$A + (A.B) = A$$
$$A.(A + B) = A$$

Annulment:	$A + 1$	$= 1$
	$A.0$	$= 0$
Association:	$(A + B) + C$	$= A + (B + C)$
	$(A.B).C$	$= A.(B.C)$
Commutation:	$A + B$	$= B + A$
	$A.B$	$= B.A$
Complements:	$A + \bar{A}$	$= 1$
	$A.\bar{A}$	$= 0$
De Morgan's:	$\overline{(A + B)}$	$= \bar{A}.\bar{B}$
	$\overline{(A.B)}$	$= \bar{A} + \bar{B}$
Distributive:	$A.(B + C)$	$= (A.B) + (A.C)$
	$A + (B.C)$	$= (A + B).(A + C)$
Double negation:	$\bar{\bar{A}}$	$= A$
Identity:	$A + O$	$= A$
	$A.1$	$= A$
Tautology:	$A.A$	$= A$
	$A + A$	$= A$

Capacitance

The capacitance of a parallel plate capacitor can be found from

$$C = \frac{0.885\, KA}{d}$$

C is in picofarads, K is the dielectric constant (air = 1). A is the area of the plate in square cm and d the thickness of the dielectric.

Calculation of overall capacitance with:
Parallel capacitors $-C = C_1 + C_2 + \cdots$
Series capacitors $-\dfrac{1}{C} = \dfrac{1}{C_1} + \dfrac{1}{C_2} + \cdots$

Characteristic impedance

$$\text{(open wire)}\ Z = 276 \log \frac{2D}{d}\ \text{ohms}$$

where

$\left. \begin{array}{l} D = \text{wire spacing} \\ d = \text{wire diameter} \end{array} \right\}$ in same units.

$$\text{(coaxial)}\ Z = \frac{138}{\sqrt{(K)}} \log \frac{d_o}{d_i}\ \text{ohms}$$

where K = dielectric constant, d_o = outside diameter of inner conductor, d_i = inside diameter of outer conductor.

Dynamic resistance

In a parallel-tuned circuit at resonance the dynamic resistance is

$$R_d = \frac{L}{Cr} = Q\omega L = \frac{Q}{\omega C} \text{ ohms}$$

where L = inductance (henries), C = capacitance (farads), r = effective series resistance (ohms). Q = Q-value of coil, and $\omega = 2\pi \times$ frequency (hertz).

Frequency – wavelength – velocity

(See also Resonance)
The velocity of propagation of a wave is

$$v = f\lambda \text{ metres per second}$$

where f = frequency (hertz) and λ = wavelength (metres).

For electromagnetic waves in free space the velocity of propagation v is approximately 3×10^8 m/sec, and if f is expressed in kilohertz and λ in metres

$$f = \frac{300\,000}{\lambda} \text{ kilohertz} \qquad f = \frac{300}{\lambda} \text{ megahertz}$$

or

$$\lambda = \frac{300\,000}{f} \text{ metres} \qquad \lambda = \frac{300}{f} \text{ metres}$$

$$f \text{ in kilohertz} \qquad f \text{ in megahertz}$$

Impedance

The impedance of a circuit comprising inductance, capacitance and resistance in series is

$$Z = \sqrt{R^2 + \left(\omega L - \frac{1}{\omega C}\right)^2}$$

where R = resistance (ohms), $\omega = 2\pi \times$ frequency (hertz). L = inductance (henries), and C = capacitance (farads).

Inductance

Single layer coils

$$L(\text{in microhenries}) = \frac{a^2 N^2}{9a + 10l} \text{ approximately}$$

If the desired inductance is known, the number of turns required may be determined by the formula

$$N = \frac{5L}{na^2} \left[1 + \sqrt{\left(1 + \frac{0.36n^2 a^3}{L} \right)} \right]$$

where N = number of turns, a = radius of coil in inches, n = number of turns per inch. L = inductance in microhenries (μH) and l = length of coil in inches.

Calculation of overall inductance with:
Series inductors $-L = L_1 + L_2 + \cdots$
Parallel inductors $-\dfrac{1}{L} = \dfrac{1}{L_1} + \dfrac{1}{L_2} + \cdots$

Meter conversions

Increasing range of ammeters or milliammeters
Current range of meter can be increased by connecting a shunt resistance across meter terminals. If R_m is the resistance of the meter; R_s the value of the shunt resistance and n the number of times it is wished to multiply the scale reading, then

$$R_s = \frac{R_m}{(n-1)}$$

Increasing range of voltmeters
Voltage range of meter can be increased by connecting resistance in series with it. If this series resistance is R_s and R_m and n as before, then $R_s = R_m \times (n-1)$.

Negative feedback

Voltage feedback

$$\text{Gain with feedback} = \frac{A}{1 + Ab}$$

where A is the original gain of the amplifier section over which feedback is applied (including the output transformer if included) and b is the fraction of the output voltage fed back.

$$\text{Distortion with feedback} = \frac{d}{1 + Ab} \text{ approximately}$$

where d is the original distortion of the amplifier.

Ohm's Law

$$I = \frac{V}{R} \quad V = IR \quad R = \frac{V}{I}$$

where I = current (amperes), V = voltage (volts), and R = resistance (ohms).

Power

In a d.c. circuit the power developed is given by

$$W = VI = \frac{V^2}{R} = I^2 R \text{ watts}$$

where V = voltage (volts), I = current (amperes), and R = resistance (ohms).

Power ratio

$$P = 10 \log \frac{P_1}{P_2}$$

where P = ratio in decibels, P_1 and P_2 are the two power levels.

Q

The Q value of an inductance is given by

$$Q = \frac{\omega L}{R}$$

Radio horizon distance

The radio horizon at VHF/UHF and up is approximately 15% further than the optical horizon. Several equations are used in calculating the distance. If D is the distance to the radio horizon, and H is the antenna height, then:

$$D = k\sqrt{H}$$

1. When D is in statute miles (5280 feet) and H in feet, then $K = 1.42$.
2. When D is in nautical miles (6000 feet) and H in feet, then $K = 1.23$.
3. When D is in kilometres and H is in metres, then $K = 4.12$.

Reactance

The reactance of an inductor and a capacitor respectively is given by

$$X_L = \omega L \text{ ohms} \qquad X_C = \frac{1}{\omega C} \text{ ohms}$$

where $\omega = 2\pi \times$ frequency (hertz), $L =$ inductance (henries), and $C =$ capacitance (farads).

The total resistance of an inductance and a capacitance in series is $X_L - X_C$.

Resistance

Calculation of overall resistance with:
Series resistors $- \text{R} = \text{R}_1 + \text{R}_2 + \cdots$

Parallel resistors $- \dfrac{1}{\text{R}} = \dfrac{1}{\text{R}_1} + \dfrac{1}{\text{R}_2} + \cdots$

Resonance

The resonant frequency of a tuned circuit is given by

$$f = \frac{1}{2\pi\sqrt{LC}} \text{ hertz}$$

where $L =$ inductance (henries), and $C =$ capacitance (farads). If L is in microhenries (μH) and C is picofarads, this becomes

$$f = \frac{10^6}{2\pi\sqrt{LC}} \text{ kilohertz}$$

The basic formula can be rearranged

$$L = \frac{1}{4\pi^2 f^2 C} \text{ henries} \qquad C = \frac{1}{4\pi^2 f L} \text{ farads}$$

Since $2\pi f$ is commonly represented by ω, these expressions can be written

$$L = \frac{1}{\omega^2 C} \text{ henries} \qquad C = \frac{1}{\omega^2 L} \text{ farads}$$

Time constant

For a combination of inductance and resistance in series the time constant (i.e. the time required for the current to reach 63% of its final value) is given by

$$\tau = \frac{L}{R} \text{ seconds}$$

where L = inductance (henries), and R = resistance (ohms).

For a combination of capacitance and resistance in series the time constant (i.e. the time required for the voltage across the capacitance to reach 63% of its final value) is given by

$$\tau = CR \text{ seconds}$$

where C = capacitance (farads), and R = resistance (ohms).

Transformer ratios

The ratio of a transformer refers to the ratio of the number of turns in one winding to the number of turns in the other winding. To avoid confusion it is always desirable to state in which sense the ratio is being expressed: e.g. the 'primary-to-secondary' ratio n_p/n_s. The turns ratio is related to the impedance ratio thus

$$\frac{n_p}{n_s} \sqrt{\frac{Z_p}{Z_s}}$$

where n_p = number of primary turns, n_s = number of secondary turns, Z_p = impedance of primary (ohms), and Z_s = impedance of secondary (ohms).

Wattage rating

If resistance and current values are known,

$$W = I^2 R \text{ when } I \text{ is in amperes}$$

334

or

$$W = \frac{\text{milliamps}^2}{1\ 000\ 000} \times R$$

If wattage rating and value of resistance are known, the safe current for the resistor can be calculated from

$$\text{milliampers} = 1.000 \times \sqrt{\frac{\text{watts}}{\text{ohms}}}$$

Wavelength of tuned circuit

Formula for the wavelength in metres of a tuned oscillatory circuit is: $1885\sqrt{LC}$, where L = inductance in microhenries and C = capacitance in microfarads.

26.9 Colour codes

26.9.1 Resistor and capacitor colour coding

Resistors | Resistors (old type) | Tubular capacitors | 'Candy stripe' capacitors

Tantalum capacitors

	1	2	3	4
Black	–	0	× 1	10 V
Brown	1	1	× 10	
Red	2	2	× 100	
Orange	3	3	–	
Yellow	4	4	–	6.3 V
Green	5	5	–	16 V
Blue	6	6	–	20 V
Violet	7	7	–	
Grey	8	8	× 0.01	25 V
White	9	9	× 0.1	3 V
			(Pink 35 V)	

Resistor and capacitor colour coding

Note that adjacent bands may be of the same colour unseparated.

Colour	Brand A	Brand B	Band C (multiplier)		Band D (tolerance)			Band E	
			Resistors	Capacitors	Resistors	Capacitors Up to 10 pF	Over 10 pF	Resistors	Polyester capacitors
Black	–	0	1	1	–	2 pF	±20%	–	–
Brown	1	1	10	10	±1%	0.1 pF	±1%	–	–
Red	2	2	100	100	±2%	–	±2%	–	250 v.w.
Orange	3	3	1 000	1 000	–	–	±2.5%	–	–
Yellow	4	4	10 000	10 000	–	–	–	–	–
Green	5	5	100 000	–	–	0.5 pF	±5%	–	–
Blue	6	6	1 000 000	–	–	–	–	–	–
Violet	7	7	10 000 000	–	–	–	–	–	–
Grey	8	8	10^8	0.01 µF	–	0.25 pF	–	–	–
White	9	9	10^9	0.1 µF	±10%	1 pF	±10%	–	–
Silver	–	–	0.01	–	±10%	–	–	–	–
Gold	–	–	0.1	–	±5%	–	–	–	–
Pink	–	–	–	–	–	–	–	Hi-Stab.	–
None	–	–	–	–	±20%	–	–	–	–

Preferred values

E12 Series

1.0 1.2 1.5 1.8 2.2 2.7
3.3 3.9 4.7 5.6 6.8 8.2
and their decades

E24 Series

1.0 1.1 1.2 1.3 1.5 1.6
1.8 2.0 2.2 2.4 2.7 3.0
3.3 3.6 3.9 4.3 4.7 5.1
5.6 6.2 6.8 7.5 8.2 9.1
and their decades

26.9.2 Resistor and capacitor letter and digit code table (BS 1852)

Resistor values are indicated as follows:

0.47 Ω	marked	R47
1 Ω		1R0
4.7 Ω		4R7
47 Ω		47R
100 Ω		100R
1 kΩ		1K0
10 kΩ		10K
10 MΩ		10M

A letter following the value shows the tolerance.

F = ±1%; G = ±2%; J = ±5%; K = ±10%; M = ±20%; R33M = 0.33Ω ± 20%; 6K8F = 6.8 kΩ ± 1%.

Capacitor values are indicated as:

0.68 pF	marked	p68
6.8 pf		6p8
1000 pF		1n0
6.8 nf		6n8
1000 nF		1μ0
6.8 μF		6μ8

Tolerance is indicated by letters as for resistors. Values up to 999 pF are marked in pF, from 1000 pf to 999 000 pF (= 999 nF) as nF (1000 pF = InF) and from 1000 nF (= 1 μF) upwards as μF.

Some capacitors are marked with a code denoting the value in pF (first two figures) followed by a multiplier as a power of ten (3 = 10^3). Letters denote tolerance as for resistors but C = ±0.25 pf. E.g. 123 J = 12 pF × 10^3 ± 5% = 12 000 pF (or 0.12 μF).

Index

338